SpringerBriefs in Materials

The SpringerBriefs Series in Materials presents highly relevant, concise monographs on a wide range of topics covering fundamental advances and new applications in the field. Areas of interest include topical information on innovative, structural and functional materials and composites as well as fundamental principles, physical properties, materials theory and design. SpringerBriefs present succinct summaries of cutting-edge research and practical applications across a wide spectrum of fields. Featuring compact volumes of 50 to 125 pages, the series covers a range of content from professional to academic. Typical topics might include

- A timely report of state-of-the art analytical techniques
- A bridge between new research results, as published in journal articles, and a contextual literature review
- A snapshot of a hot or emerging topic
- An in-depth case study or clinical example
- A presentation of core concepts that students must understand in order to make independent contributions

Briefs are characterized by fast, global electronic dissemination, standard publishing contracts, standardized manuscript preparation and formatting guidelines, and expedited production schedules.

More information about this series at http://www.springer.com/series/10111

Vijaykumar V. Jadhav ·
Rajaram S. Mane · Pritamkumar V. Shinde

Bismuth-Ferrite-Based Electrochemical Supercapacitors

 Springer

Vijaykumar V. Jadhav
Department of Physics
Shivaji Mahavidyalaya
Udgir, Maharashtra, India

Department of Material Science
and Engineering
Guangdong Technion, Israel Institute
of Technology
Shantou, China

Department of Material Science
and Engineering
Technion-Israel Institute
of Technology
Haifa, Israel

Pritamkumar V. Shinde
Department of Material Science
and Engineering, Global Frontier
R&D Center
Pusan National University
Busan, Korea (Republic of)

Rajaram S. Mane
School of Physical Sciences
Swami Ramanand Teerth Marathwada
University
Nanded, Maharashtra, India

ISSN 2192-1091   ISSN 2192-1105   (electronic)
SpringerBriefs in Materials
ISBN 978-3-030-16717-2   ISBN 978-3-030-16718-9   (eBook)
https://doi.org/10.1007/978-3-030-16718-9

This Springer imprint is published by the registered company Springer Nature Switzerland AG
The registered company address is: Gewerbestrasse 11, 6330 Cham, Switzerland

Preface

This book provides the significance of supercapacitors in addition to batteries and fuel cells. Supercapacitor is the prominent and alternative energy storage device to lithium-ion battery in near future. This book also provides information about in-depth history, types, designing processes, operation mechanisms, and advantages and disadvantages of supercapacitors. The soul of supercapacitors is electrolytes which are mostly ignored character during investigations, are also briefed, as the fabrication and design of supercapacitors are essentially important. Bismuth ferrite for supercapacitor applications provides a snapshot of the present status of this rapidly growing field.

The main important concept of this book is to explain the use of bismuth ferrite electrode material in supercapacitors. It also provides a much-needed, up-to-date overview of different kinds of structural arrangement bismuth ferrite-based systems follow, with a focus on their properties, synthesis methods, and applications as electrochemical supercapacitors. It introduces readers to the basic structure and properties of ferrites, in general, focusing on the selection criteria for ferrite materials in electrochemical energy storage applications. Along with coverage of ferrite synthesis methods, it discusses bismuth ferrite structures in unary, binary, and mixed ferrite nanostructure systems, as well as future perspectives and limitations for using ferrites in electrochemical supercapacitors. Finally, the market status of supercapacitors and a discussion pointing out the future scope and directions of next-generation bismuth ferrite-based supercapacitors are highlighted, making this a comprehensive book on the latest, cutting-edge research in the field.

Udgir, India/Shantou, China/Haifa, Israel   Vijaykumar V. Jadhav
Nanded, India   Rajaram S. Mane
Busan, Korea (Republic of)   Pritamkumar V. Shinde

Contents

About the Authors

Vijaykumar V. Jadhav is an Assistant Professor at the Department of Physics, Shivaji Mahavidyalaya, Udgir, Maharashtra, India. He joined GTIIT in 2019 as a research scientist at Material Science and Engineering Laboratory, Guangdong Technion, Israel Institute of Technology, Shantou, China. His research is focused on the synthesis of metal oxide and MXene materials for energy storage and generation technologies. He is a recipient of a Government of Ireland Postdoctoral Fellowship in 2016 to develop advanced materials for 3D lithium–ion batteries. Based on 3D-printed rechargeable nickel–zinc battery research, he has been awarded Young Scientist through Lindau Nobel Laureate Meeting, Lindau, Germany. He is an active reviewer for ACS, Elsevier, RSC, and Springer journals. He has published 40 peer-reviewed journal articles and three chapters. He so far has delivered three invited talks and two oral presentations at prestigious scientific peer conferences with international acclaims and awards, generated research funds in excess of >€ 350,000, supervised students/junior researchers, and actively participated in outreach and scientific dissemination for the service of wider community. He has several collaborations with eminent scientists across the globe including Australia, Saudi Arabia, Korea, China, Ireland, Poland, and India. He also has collaborations and interactions with over 25 junior and senior colleagues in India, South Korea, China, and Ireland with whom he has coauthored his publications. As an asset, he successfully qualified the national eligibility test (NET) with 364th rank, conducted by CSIR, in Physics, New Delhi, India, which helped him to get Ph.D. admission which is also one of the essential eligibility criteria for the assistant professor position in various national universities in India. He has demonstrated outstanding ability as independent research, great teamwork, and international network in six countries of three continents (Asia, Europe, and Australia). Embracing several countries and cultures is important in someone, looking to dedicate a life to exploration in science and technology.

Prof. Rajaram S. Mane is currently working as a Professor at School of Physical Science, S.R.T.M. University, Nanded, Maharashtra, India. Presently, he is also a director of Innovation, Incubation and Linkages. He has received his Ph.D. in Physics from the Shivaji University, Kolhapur, India, in 2000 and worked as a postdoctoral fellow at Hanyang University, Korea. He also was on the research faculty of Oxford University, Oxford, UK. Since 2010, he has been a regular professor at S.R.T.M., University, Nanded, India, and a Visiting Professor of Pusan National University, Korea. Few technology transfers, book chapters, review articles and awards like Brain Pool Fellow are on his credit. With more than 331 peer reviewed research journal articles, 8700 citations, and 192 i10 indexes, he is actively engaged in synthesis of 2D and 3D metal oxides/chalcogenides/carbides/ nitrides/fluorides for solar cells, electrochemical supercapacitors, chemical sensors, and bioactive applications. His major interests include synthesis of novel nanostructures for energy conversion and storage device technologies.

Pritamkumar V. Shinde is a Postdoctoral Researcher at Material Science and Engineering Department, Pusan National University, Busan, South Korea. He has published 30 peer-reviewed journal articles and two chapters. He has received his Ph.D. in chemistry from Bharati Vidyapeeth Deemed University, Pune, India.

Chapter 1
Introduction

1.1 Introduction

Significant shift in the global climate and the shortage of fossil fuels require the society to acquire green energy. Energy is one of the most essential aspects in the twenty-first century which is essential to human lives. Energy production and consumption, a function of combustion of fossil fuel, is going to disturb the world economy. So, there has been an increasing demand for environmentally friendly, high-performance, and cost-effective renewable energy storage devices [1, 2]. Since 1980, electrochemical supercapacitors (ESs) have gained significant attention due to their high-power density, longer cycle life than batteries, and higher energy density than conventional dielectric capacitors. Such outstanding advantages have made them potential alternatives for hybrid electric vehicles, digital cameras, industrial equipments, and other renewable energy storage devices. Also, due to a fast-growing market for portable electronic devices, such as mobile phones, notebooks, laptops, computers, and their development trend of being small, lightweight, and flexible, have brought about an ever-rising and urgent demand for eco-friendly electrochemical energy storage and conversion systems that are lying between capacitors and the batteries [3–8]. In order to mitigate the serious worries concerning energy disaster, numerous substitute technologies have come up into reality [5]. In upcoming days, ESs will be painstaking as a significant model in technology due to the acceptance of energy storage systems [3]. Nowadays, ESs can be a good alternative to batteries which offer raised power densities, density, and stable, long discharge time [7]. Figure 1.1 endows the functional position between batteries and predictable capacitors obtained by blend of a high-power capability, coupled with good specific energy. The ESs hit the apex coming to the power density feature but have considerably lower power density compared to conventional capacitors displayed in Ragone plot for different energy storage devices [9]. The predictable capacitor involves the dielectric plates for electrostatic charge storage, whereas supercapacitor (hybrid) comprises submerged electrodes within electrolyte solution kept apart through a separator prospering electrolytic ions diffusion but hinders the direct contact of electrodes to avoid short circuiting [10]. Hybrid

© The Authors 2020
V. V. Jadhav et al., *Bismuth-Ferrite-Based Electrochemical Supercapacitors*,
SpringerBriefs in Materials, https://doi.org/10.1007/978-3-030-16718-9_1

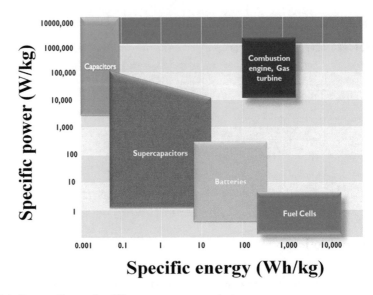

Fig. 1.1 Ragone diagram for different energy storage devices

supercapacitors may be asymmetric or symmetric based on either two different electrodes [11]. In comparison to present rechargeable batteries, the ESs include EDLC,
pseudocapacitor, and hybrid supercapacitor that furnish much greater power density
owing to a characteristic of reactions experienced on the exterior layer of the electrode materials to store the charge [12]. The present rechargeable batteries mainly
depend on intercalation and de-intercalation of cations controlled by diffusion that
restricts their charging and discharging rates or power density [13]. The hybrid ESs
possess the ability to store a huge quantity of charge furnished at elevated power
rates in comparison to rechargeable batteries [14]. Hence, hybrid ESs promise to be
a reciprocal choice compared to rechargeable batteries urging to high-power delivery
or fast energy yield [15]. In contradiction of batteries, the charge restricts their energy
density to an inferior value; however, hybrid ESs can be the better choice in which the
energy density comes out ahead due to active material-specific capacitance and net
cell voltage [16, 17]. The hybrid ES systems comprising of non-aqueous redox materials are being the advanced trend in recent times that are enthusiastically investigated
and extensively developed [18–24]. Conversely, the expansion of hybrid supercapacitor proficiently by the potential gap between the two different electrodes increases
the overall cell voltage that cannot be negotiated [25]. Therefore, an approach of
amplifying specific capacitance (SC) in addition to cell voltage and enhancing the
performance is required. The case of increasing the SC is concerned with EDLC and
pseudocapacitive electrodes, while the latter approach of increasing cell voltage is
associated with hybrid supercapacitors [26]. This leads to a considerable betterment
in performance of respective applications. The other method involves the growth of
nanosized electrode materials to diminish the diffusion length and most importantly

to provide a higher outer layer area [27]. This method is mainly anxious with super-capacitors and rechargeable batteries. The important factor is the recharge time of hybrid supercapacitors than the conventional lead–acid battery and other recharge-able batteries [28]. Furthermore, ESs signify a new class of technology among the other energy storage devices which store larger amount of energy than traditional capacitors and deliver more power than batteries. ESs can also be regarded as func-tioning like rechargeable batteries in storing or delivering electric charge, and their mechanism of charge storage differs from that typically operating in batteries.

1.1.1 General Information of Energy Storage Devices

The main difference between the supercapacitors, batteries, and fuel cells is that supercapacitors and batteries store energy, while a fuel cell generates energy by con-verting available fuel. A fuel cell can have supercapacitor or battery as a system com-ponent to store the electricity its generating. Batteries, supercapacitors, and fuel cells consist of two electrode systems in contact with an electrolyte solution. Figure 1.2 shows the basic operation mechanism of battery, fuel cell, and supercapacitor.

1.1.1.1 Battery

A battery is a device consisting of one or more electrochemical cells with two ter-minals; its positive terminal is the cathode, and negative terminal is the anode with external connection provided to the power devices such as smartphone and electric items [29, 30]. When battery is connected to the external load, the electron crosses from negative terminal to the positive terminal by creating and electric current. This current may power to the light bulb, a clock, a cell phone, and other electronic devices. This electrical energy is generated by conversion of chemical energy through redox reactions at the anode and cathode via electrolyte. The anode is at lower potential than cathode. Batteries are closed system, and charge storage occurs at anode and cathode,

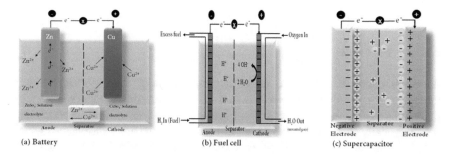

(a) Battery   (b) Fuel cell   (c) Supercapacitor

Fig. 1.2 Basic operation mechanisms of; (**a**) battery, (**b**) fuel cell, and (**c**) supercapacitor devices

thereby the energy storage and conversion occur in the same cell. Batteries are classified into two types, i.e., primary and secondary batteries. The batteries which cannot be recharged after used are known as primary batteries. These are disposable batteries. Because chemical reaction takes place in primary batteries cannot be reversed, and used active materials do not go back to their original forms. These batteries normally used in portable devices with minimal current drain like alkaline batteries and zinc–carbon batteries. The rechargeable batteries are referred as secondary batteries which are built to be recharged and reused many times with help of an electric current. The composition of the electrodes can be restored by reverse current. The lead–acid batteries usually applied in vehicles are the prehistoric example of the rechargeable batteries. Another type of rechargeable batteries is called as dry batteries. These are useful in portable devices such as laptop and mobile phones, e.g., lithium-ion (Li-ion), nickel–metal hydride (NiMH), and nickel–cadmium (NiCd) cells and lead batteries. Therefore, the battery is a more practical option than fuel cell, but charge-and-discharge process in batteries is a slow process and can degrade the chemical compounds inside the battery over time and owing to material damage, and batteries possess low power density and lose ability to retain energy all the way through their lifetime.

Limitations:

Li-ion	NiMH	NiCd	Lead
Requires protection circuit	Limited service life	Relatively low energy density	Low energy density
Moderate discharge current	Limited discharge current	Memory effect	Limited discharge cycles
Expensive to manufacture	High self-discharge	Environmentally unfriendly	Environmentally unfriendly
Not fully matured	High maintenance	Needs recharging after storage	Transportation restrictions
Lower energy density and decreased cycle count	More expensive		Thermal runaway can occur with improper charging
			Cannot be stored in a discharged condition

1.1.1.2   Fuel Cells

An electrochemical cell which converts chemical energy into electrical energy through redox reaction is referred as fuel cell. The reactions produce electricity taking place at the electrode structure. Fuel cell has also an electrolyte which carries electrically charged particles from one electrode to the other and catalyst which speeds up the reaction at the electrode surface. There are several kinds of fuel cells which are classified by the type of electrolyte used. In fuel cell, hydrogen enters at the anode where a catalyst causes the fuel which undergoes oxidation reaction and generates ions and electrons. The ionized hydrogen atoms carry a positive electrical charge which moves from anode to cathode through the electrolyte. Instantaneously, electrons flow from the anode to the cathode through an external circuit which produces direct electric current. At the cathode, another catalyst causes ions, electrons, and oxygen to react, forming water and other possible products. Operation of fuel cell is different than battery. It requires continuous source of fuel and oxygen from air to sustain chemical reaction. It produces electricity continuously as long as fuel and oxygen are supplied. Individual fuel cell produces relatively small electrical potential about 0.7 V. Due to this reason, cells are stacked or connected in series to create sufficient voltage to meet an application requirement. Fuel cells are commonly used to power the vehicles, automobiles, buses, boats, and motorcycles. Fuel cells are classified on the basis of electrolyte they use such as proton-exchange membrane fuel cells, solid oxide fuel cells, phosphoric acid fuel cell, solid acid fuel cell, alkaline fuel cell, and high-temperature fuel cells.

Limitations:

- The issue of hydrogen storage is still existing which makes transportation of hydrogen fuel challenging so the technologies prefer to provide alternative fuel storage and delivery methods.
- MCFCs and PAFCs, reform natural gas, providing the perfect solution for industrial use but due to large size difficult for transport and home use. SOFs also can internally reform natural gas, but it will take years to come in practice.
- The PEMs and AFCs can use fuel reformers to convert hydrocarbons, but with low efficiency, it releases small amount of pollutants.
- The work on fuel storage and conversion solutions is going on but will take years to come in practical use.

Hurdles:

- the hydrogen problem—basically H_2 is the only reasonable fuel in the estimative future,
- difficulty in hydrogen storage and supply infrastructure,
- the effectiveness of the hydrogen from fossil fuel is still questionable,
- need to maintain temperature, and
- high cost.

1.1.1.3 Electrochemical Supercapacitors

The electrochemical supercapacitor, an electrochemical device which stores the charge on electrode with high capacitance, is also referred as ultra/hybrid capacitor together with EDLCs which have anode and cathode as negative electrode and working electrode, respectively, by means of electrolyte and separator in between them. The electrolyte may in the form of solid, liquid, and gel. The charge–discharge processes of supercapacitor are much faster than rechargeable batteries. On the basis of their configuration, supercapacitors are referred as symmetric and asymmetric in which energy storage depends on the accumulation of charge or reversible redox reactions [1, 2]. Long discharging time, easy operating, and high temperature range are the advantages of supercapacitor over ordinary batteries and fuel cells [31]. The hybrid supercapacitors are the devices have high capacitance and energy storage capabilities and have the combining the properties of EDLCs and pseudocapacitors, i.e., combination of EDLC and pseudocapacitor [32–34], whereas the storage of energy is attained due to rapid repeatable redox reactions among electroactive units lying on active electrode material and an electrolyte solution in pseudocapacitor [35]. One-half of the hybrid supercapacitor acts as EDLC, while another half behaves as pseudocapacitor [36]. Hence, we focus on supercapacitor and supercapattery devices, and the associated nanostructured, redox active, and semiconductor materials such as electronically conducting transition metal oxides (TMOs) and their ferrites, especially bismuth ferrites. As a new ESs device, supercapattery aims to achieve comparable performance to supercapacitor in power capability and cycle life, and to battery in energy capacity.

1.1.2 *Ferrites*

Transition-metal ferrites with a spinel structure (MFe_2O_4 with $M = Co^{2+}, Ni^{2+}, Cu^{2+}$, Zn^{2+}, etc.) are used in many technological applications such as magnetic memory devices [37–47]. Cobalt ferrite ($CoFe_2O_4$) is an important member of magnetic ferrite class which can be easily accessible at room temperature [47, 48]. Nickel ferrite ($NiFe_2O_4$) has also been intensively studied over the past several decades because of its high magnetization and low coercively, making it promising for applications requiring the use of soft-magnetic materials [49, 50]. Both crystalline $NiFe_2O_4$ and $CoFe_2O_4$ have the space group Fd3m, known as inverse spinel ferrites in which octahedral sites are occupied by Co^{2+}/Ni^{2+} divalent ions, while an equal number of Fe^{3+} ions reside on tetrahedral and octahedral sites [51–55]. It has been shown, for example, that enhancement in magnetization can be achieved with decreasing grain (or particle, size) but also with better defined crystallinity, and material characteristics are inherently difficult to reconcile [56, 57]. Overall, it has been shown in several studies that the cyclability, including active material utilization, kinetics, and energy efficiency, can significantly be improved with low-dimensional porous nanostructures, which are capable of mitigating, to some extent, the problem of particle fracture due to volume change (electrode breathing) during lithiation/delithiation and of reducing the average diffusion distance [58–64]. Besides, in situ control of magnetism through lithium insertion reaction has currently drawn increasing interest in the battery and magnetoelectrics community [65–71]. For some nanocrystalline spinel ferrites such as $CuFe_2O_4$ and $ZnFe_2O_4$, it has been demonstrating that a bulky and revocable variations in magnetization can be attained by monitoring cautiously the volume of inserted lithium (to prevent irreversible structural changes) [67]. Following the same line, the general idea of combining magnetism and LIB energy storage concepts has been extended to transition-metal ferrites as thin films with ordered mesoporous morphology. To investigate the magnetic material at room temperature, the cubic mesostructured thin films can be prepared by an evaporation-induced self-assembly (EISA) method together with $CoFe_2O_4$, $NiFe_2O_4$, and $Co_{0.5}Ni_{0.5}Fe_2O_4$ [72–74]. It was earlier reported that the oxides can be entirely crystallized, whereas protecting the nanoscale properties allow to modify the magnetic behavior at room temperature over facile insertion/extraction of lithium into the spinel lattice. However, referring to the commercial aspect and good capacitance, high rate performance with the maximum working potential range, less toxicity with low cost as essential and thus, the cost-effective electrodes essential throughout the electrochemical investigations. In this context, the use of ferrites for supercapacitor applications can be a good option. An auspicious approach to improve the performance is scheming ternary ferrite-based hybrids. Spinel ferrite may also consist of a mixture of two divalent metal ions; nonetheless, the cation distribution of such ferrites can affect the surface properties of the spinel ferrites, which makes them catalytically active. Because of their small size and large number of cations, for coordination sites, nanocrystallites can easily boost the chemical reactions. Moreover, the bismuth oxide (Bi_2O_3), one of the important transition metal oxides, is of low cost, relatively abundant, and environmentally friendly [75]. As well as five polymorphs viz., α, β, γ, δ, and ε in different nanostructures its non-toxicity, high refractive index, and magnetic etc., devices [75–81]. To date, synthesis of Bi_2O_3 in different polymorphs and nanostructures is preferred through various solution processes likewise electrodeposition, solvothermal, chemical vapor deposition, atomic layer deposition, chemical bath deposition, successive ionic layer adsorption and reaction, spray pyrolysis and microwave-assisted synthesis method, etc., [82–91] where, in most of the cases, either plain or single morphology can be obtained and moreover, synthesis time is sometimes highly prolonged.

1.2 Conclusions

In this chapter, the general stragegy used while storing electrical energy through electrochemical devices likes supercapacitors, batteries, and fuel cells, and their limitations, and hurdles in operation are briefed. An importance of supercapacitors over batteres, and fuel cells as significant energy storage devices has also been highlighted. Supercapacitor demonstrate high energy, and power density useful for consumer electronic devices, power backups on replacing the batteries, and also in electrical vehicles. Supercapacitor can also be utilized as rechargeable batteries to provide additional power. In order to fabricate the energy storage devices, working electrode is essential with high surface area and conductivity. Ferrites can be one of the important family of electrode materials suitable for energy storage applications.

References

1. K. Poonam, A. Sharma, S.K. Arora, J. Tripathi, Energy Storage. **21**, 801–825 (2019)
2. A. Muzaffar, M.B. Ahamed, K. Deshmukh, J. Thirumalai, Renew Sust Energ Rev. **101**, 123–145 (2019)
3. P. Simon, Y. Gogotsi, Nature Mater. **7**, 845–854 (2008)
4. C. Choi, D.S. Ashby, D.M. Butts, R.H. DeBlock, Q. Wei, J. Lau, B. Dunn, Nat. Rev. Mater. (2019). https://doi.org/10.1038/s41578-019-0142-z
5. L.L. Zhang, X.S. Zhao, Chem. Soc. Rev. **38**, 2520–2531 (2009)
6. C. Liu, F. Li, L.P. Ma, H.M. Cheng, Adv. Mater. **22**, E28–E62 (2010)
7. J.R. Miller, Electrochim. Acta **52**, 1703–1708 (2006)
8. N.W. Duffy, W. Baldsing, A.G. Pandolfo, Electrochim. Acta **54**, 535 (2008)
9. G. Wang, L. Zhang, J. Zhang, Chem. Soc. Rev. **41**, 797–828 (2012)
10. X. Wang, Y.L. Yang, R. Fan, Y. Wang, Z.H. Jiang, J. Alloy. Compd. **504**, 32–36 (2010)
11. P.C. Chen, G. Shen, Y. Shi, H. Chen, C. Zhou, ACS Nano **4**, 4403–4411 (2010)
12. S.W. Lee, J. Kim, S. Chen, P.T. Hammond, Y. Shao-Horn, ACS Nano **4**, 3889–3896 (2010)
13. D. Zhang, H. Yan, Y. Lu, K. Qiu, C. Wang, C. Tang, Y. Zhang, C. Cheng, Y. Luo, Nanoscale Res. Lett. **9**, 139 (2014)
14. H.W. Wang, Z.A. Hu, Y.Q. Chang, Y.L. Chen, H.Y. Wu, Z.Y. Zhang, Y.Y. Yang, J. Mater. Chem. **21**, 10504–10511 (2011)
15. A.G. Pandolfo, A.F. Hollenkamp, J. Power Sources **157**, 11–27 (2006)
16. I. Plitz, A. DuPasquier, F. Badway, J. Gural, N. Pereira, A. Gmitter, Appl. Phys. A Mater. Sci. Process. **82**, 615–626 (2006)
17. Y. Li, P. Hasin, Y. Wu, Adv. Mater. **22**, 1926–1929 (2010)
18. H. Chai, X. Chen, D. Jia, W. Zhou, Rare Met. **30**, 85–89 (2011)
19. K. Naoi, Fuel Cells **105**, 825–833 (2010)
20. S. Huang, Z. Wen, X. Zhu, Z. Gu, Electrochem. Commun. **6**, 1093–1097 (2004)
21. L. Luo, J. Wu, J. Xu, V.P. Dravid, NUANCE Center. Microsc. Microanal. **20**, 1618–1619 (2014)
22. M. Tachibana, T. Ohishi, Y. Tsukada, A. Kitajima, H. Yamagishi, M. Murakami, Electrochemistry **79**, 882–886 (2011)
23. J. Yan, Z. Fan, W. Sun, G. Ning, T. Wei, Q. Zhang, R. Zhang, L. Zhi, F. Wei, Adv. Func. Mater. **22**, 2632–2641 (2012)
24. V. Gupta, S. Gupta, N. Miura, J. Power Sources **175**, 680–685 (2008)
25. Z. Fan, J. Yan, T. Wei, L. Zhi, G. Ning, T. Li, F. Wei, Adv. Funct. Mater. **21**, 2366–2375 (2011)
26. J. Li, N. Wang, J. Tian, W. Qian, W. Chu, Adv. Funct. Mater. **28**, 1806153 (2018)
27. C. Xiang, Y. Liu, Y. Yin, P. Huang, Y. Zou, M. Fehse, Z. She, F. Xu, D. Banerjee, D.H. Merino, A. Longo, H.B. Kraatz, D.F. Brougham, B. Wu, L. Sun, A.C.S. Appl, Energy Mater. **2**, 3389–3399 (2019)
28. Y. Jiao, C. Qu, B. Zhao, Z. Liang, H. Chang, S. Kumar, R. Zou, M. Liu, K.S. Walton, A.C.S. Appl, Energy Mater. **2**, 5029–5038 (2019)
29. M. Winter, R.J. Brodd, Chem. Rev. **104**(10), 4245–4270 (2004)
30. A. Manthiram, ACS Cent. Sci. **3**, 1063–1069 (2017)
31. X. Liu, O.O. Taiwo, C. Yin, M. Ouyang, R. Chowdhury, B. Wang, H. Wang, B. Wu, N.P. Brandon, Q. Wang, S.J. Cooper, Adv. Sci. **6**, 1801337 (2019)
32. S. Wu, Y. Chen, T. Jiao, J. Zhou, J. Cheng, B. Liu, S. Yang, K. Zhang, W. Zhang, Adv. Energy Mater. **9**, 1902915 (2019)
33. E. Lim, H. Kim, C. Jo, J. Chun, K. Ku, S. Kim, H. Ik Lee, I. S. Nam, S. Yoon, K. Kang, J. Lee, ACS Nano **8** 8968–8978 (2014)
34. Z. Lu, Z. Chang, W. Zhu, X. Sun, Chem. Commun. **47**, 9651–9653 (2011)
35. X. Chen, Y. Zhu, M. Zhang, J. Sui, W. Peng, Y. Li, G. Zhang, F. Zhang, X. Fan, ACS Nano **13**, 9449–9456 (2019)
36. D.P. Dubal, O. Ayyad, V. Ruiz, P. Gomez-Romero, Chem. Soc. Rev. **44**, 1777–1790 (2015)
37. N.A. Frey, S. Peng, K. Cheng, S. Sun, Chem. Soc. Rev. **38**(2009), 2532–2542 (2009)

38. W. Hu, N. Qin, G. Wu, Y. Lin, S. Li, D. Bao, J. Am. Chem. Soc. **134**, 14658–14661 (2012)
39. Q. Dai, K. Patel, G. Donatelli, S. Ren, Angew. Chem. Int. Ed. **55**, 10439–10443 (2016)
40. S.B. Han, T.B. Kang, O.S. Joo, K.D. Jung **81**, 623–628 (2007)
41. V. Polshettiwar, R. Luque, A. Fihri, H. Zhu, M. Bouhrara, J.M. Basset, Chem. Rev. **111**, 3036–3075 (2011)
42. C. Duanmu, I. Saha, Y. Zheng, B.M. Goodson, Y. Gao, Chem. Mater. **18**, 5973–5981 (2006)
43. H.M. Fan, J.B. Yi, Y. Yang, K.W. Kho, H.R. Tan, Z.X. Shen, J. Ding, X.W. Sun, M.C. Olivo, Y.P. Feng, ACS Nano **3**, 2798–2808 (2009)
44. H. Zhang, L. Li, X.L. Liu, J. Jiao, C.T. Ng, J.B. Yi, Y.E. Luo, B.H. Bay, L.Y. Zhao, M.L. Peng, N. Gu, H.M. Fan, ACS Nano **11**, 3614–3631 (2017)
45. A. H. Lu, E.L. Salabas, F. Schüth, Angew. Chem. Int. Ed. **46**, 1222–1244 (2007)
46. N.T.K. Thanh, L.A.W. Green, Nano Today **5**, 213–230 (2010)
47. Q. Dai, D. Berman, K. Virwani, J. Frommer, P.O. Jubert, M. Lam, T. Topuria, W. Imaino, A. Nelson, Nano Lett. **10**, 3216–3221 (2010)
48. T.E. Quickel, V.H. Le, T. Brezesinski, S.H. Tolbert, Nano Lett. **10**, 2982–2988 (2010)
49. K.A. Pettigrew, J.W. Long, E.F. Carpenter, C.C. Baker, J.C. Lytle, C.N. Chervin, M.S. Logan, R.M. Stroud, D.R. Rolison, ACS Nano **2**, 784–790 (2008)
50. J. Zhang, J. Fu, G. Tan, F. Li, C. Luo, J. Zhao, E. Xie, D. Xue, H. Zhang, N.J. Mellors, Y. Peng, Nanoscale **4**, 2754–2759 (2012)
51. J. Jacob, M.A. Khadar, J. Appl. Phys. **107**, 114310 (2010)
52. C. Reitz, C. Suchomski, J. Haetge, T. Leichtweiss, Z. Jagličić, I. Djerdj, T. Brezesinski, Chem. Commun. 48 (2012) 4471 – 4473
53. C. Reitz, C. Suchomski, V.S.K. Chakravadhanula, I. Djerdj, Z. Jagličić, T. Brezesinski, Inorg. Chem. **52**, 3744–3754 (2013)
54. K. Vasundhara, S.N. Achary, S.K. Deshpande, P.D. Babu, S.S. Meena, A.K. Tyagi, J. Appl. Phys. **113**, 194101 (2013)
55. C.N. Chinnasamy, A. Narayanasamy, N. Ponpandian, K. Chattopadhyay, H. Guérault, J.M. Greneche, J. Phys.: Condens. Matter. **12**, 7795–7805 (2000)
56. Y.Q. Chu, Z.W. Fu, Q.Z. Qin, Electrochim. Acta **49**, 4915–4921 (2004)
57. P. Lavela, J.L. Tirado, J. Power Sources **172**, 379–387 (2007)
58. J. Cabana, L. Monconduit, D. Larcher, M.R. Palacín, Adv. Mater. **22**, E170–E192 (2010)
59. D. Bresser, E. Paillard, R. Kloepsch, S. Krueger, M. Fiedler, R. Schmitz, D. Baither, M. Winter, S. Passerini, Adv. Energy Mater. **3**, 513–523 (2013)
60. J.M. Won, S.H. Choi, Y.J. Hong, Y.N. Ko, Y.C. Kang, Sci. Rep. **4**, 5857 (2014)
61. C. Suchomski, B. Breitung, R. Witte, M. Knapp, S. Bauer, T. Baumbach, C. Reitz, T. Brezesinski, Beilstein J. Nanotechnol. **7**, 1350–1360 (2016)
62. G. Zeng, N. Shi, M. Hess, X. Chen, W. Cheng, T. Fan, M.A. Niederberger, ACS Nano **9**, 4227–4235 (2015)
63. Z. Zhang, W. Li, R. Zou, W. Kang, Y.S. Chui, M.F. Yuen, C.S. Lee, W. Zhang, J. Mater. Chem. A **3**, 6990–6997 (2015)
64. Q.Q. Xiong, J.P. Tu, S.J. Shi, X.Y. Liu, X.L. Wang, C.D. Gu, J. Power Sources **256**, 153–159 (2014)
65. S. Dasgupta, B. Das, M. Knapp, R.A. Brand, H. Ehrenberg, R. Kruk, H. Hahn, Adv. Mater. **26**, 4639–4644 (2014)
66. T. Tsuchiya, K. Terabe, M. Ochi, T. Higuchi, M. Osada, Y. Yamashita, S. Ueda, M. Aono, ACS Nano **10**, 1655–1661 (2016)
67. S. Dasgupta, B. Das, Q. Li, D. Wang, T.T. Baby, S. Indris, M. Knapp, H. Ehrenberg, K. Fink, R. Kruk, H. Hahn, Adv. Funct. Mater. **26**, 7507–7515 (2016)
68. C. Reitz, C. Suchomski, D. Wang, H. Hahn, T. Brezesinski, J. Mater. Chem. C **4**, 8889–8896 (2016)
69. Q. Zhang, X. Luo, L. Wang, L. Zhang, B. Khalid, J. Gong, H. Wu, Nano Lett. **16**, 583–587 (2016)
70. G. Wei, L. Wei, D. Wang, Y. Tian, Y. Chen, S. Yan, L. Mei, J. Jiao, RSC Adv. **7**, 2644–2649 (2017)

71. G. Wei, L. Wei, D. Wang, Y. Chen, Y. Tian, S. Yan, L. Mei, J. Jiao, Sci. Rep. **7**, 12554 (2017)
72. C.J. Brinker, Y. Lu, A. Sellinger, H. Fan, Adv. Mater. **11**, 579–585 (1999)
73. Y. Lu, R. Ganguli, C.A. Drewien, M.T. Anderson, C.J. Brinker, W. Gong, Y. Guo, H. Soyez, B. Dunn, M.H. Huang, J.I. Zink, Nature **389**, 364–368 (1997)
74. D. Grosso, F. Cagnol, G.J.D. A.A. Soler-Illia, E.L. Crepaldi, H. Amenitsch, A. Brunet-Bruneau, A. Bourgeois, C. Sanchez, Adv. Funct. Mater. **14**, (2004) 309–322
75. T.P. Gujar, V.R. Shinde, C.D. Lokhande, R.S. Mane, S.H. Han, Appl. Surf. Sci. **250**, 161–167 (2005)
76. T. Takeyama, N. Takahashi, T. Nakamura, S. Itoh, J. Cryt. Growth **275**, 460–466 (2005)
77. W.D. He, W. Qin, X.H. Wu, X.B. Ding, L. Chen, Z.H. Jiang, Thin Solid Films **515**, 5362–5365 (2007)
78. I.I. Oprea, H. Hesse, K. Betzler, Opt. Mater. **26**, 235–237 (2004)
79. T.A. Hanna, Coord. Chem. Rev. **248**, 429–440 (2004)
80. A.M. Azad, S. Larose, S.A. Akbar, J. Mater. Sci. **29**, 4135–4151 (1994)
81. N.M. Shinde, Q.X. Xia, J.M. Yun, S. Singh, R.S. Mane, K.H. Kim, Dalton Trans. **46**, 6601–6611 (2017)
82. Z.N. Adamian, H.V. Abovian, V.M. Aroutiounian, Sens. Actuators, B **35**, 241–243 (1996)
83. E.T. Salim, Y. Al-Douri, M.S. Al Wazny, M.A. Fakhri, Sol. Energy **107**, 523–529 (2014)
84. L. Li, Y.W. Yang, G.H. Li, L.D. Zhang, Small **2**, 548–553 (2006)
85. L. Kumari, J.H. Lin, Y.R. Ma, J. Phys.: Condens. Matter **18**, 295605–295612 (2007)
86. L. Kumari, J.H. Lin, Y.R. Ma, J. Phys.: Condens. Matter **19**, 406204–406215 (2007)
87. H. Nguyen, S.A. El-Safty, J. Phys. Chem. C **115**, 8466–8474 (2011)
88. T.P. Gujar, V.R. Shinde, C.D. Lokhande, S.H. Han, Mater. Sci. Eng., B **133**, 177–180 (2006)
89. P.V. Shinde, B.G. Ghule, N.M. Shinde, Q.X. Xia, S.F. Shaikh, A.V. Sarode, R.S. Mane, K.H. Kim, New J. Chem. **42**, 12530–12538 (2018)
90. R.C. Ambare, P. Shinde, U.T. Nakate, B.J. Lokhande, R.S. Mane, Appl. Surf. Sci. **453**, 214–219 (2018)
91. P.V. Shinde, B.G. Ghule, S.F. Shaikh, N.M. Shinde, S.S. Sangale, V.V. Jadhav, S.Y. Yoon, K.H. Kim, R.S. Mane, J. Alloy. Compd. **802**, 244–251 (2019)

Chapter 2
Electrochemical Supercapacitors: History, Types, Designing Processes, Operation Mechanisms, and Advantages and Disadvantages

2.1 History

The perception of electrochemical supercapacitors (ESs) depended on the electric double-layer (EDL) existing at the interface between a conductor and its contacting electrolyte solution. The electric double-layer theory was the first proposed by Hermann von Helmholtz in 1853 and further developed by Gouy, Chapman, Grahame, and Stern [1]. On behalf of origination of double-layer theory, numerous theories and technologies, viz., supercapacitors, batteries, and fuel cells, get invented. Electrostatic and electrolytic capacitors are considered to be the first- and second-generation capacitors, respectively. With the rapid developments in materials, the third-generation capacitor known as the supercapacitor was invented [2–5]. As discussed earlier, ESs are also named as supercapacitors, ultracapacitors, or electric double-layer capacitors (EDLCs). The term 'supercapacitor' finds itself in common usage, being the trade name of the first commercial devices made by Nippon Electric Company (NEC) in 1971. The term 'ultracapacitor' has also been originated from devices made by the Pinnacle Research Institute (PRI) with low internal resistance invented in 1982 for the military applications. In 1957, Becker invented low-voltage electrolytic capacitor with porous carbon electrode and filed the patent as shown in Fig. 2.1 who used a high specific surface area-coated carbon on a metallic current collector in a sulfuric acid solution [6, 7]. The researcher from Standard Oil of Ohio, R. A. Rightmire, invented a device that stored energy in the double-layer interface in 1966 [8]. In 1971, Nippon Electronic Company (NEC) from Japan developed aqueous-electrolyte capacitors under the energy company Standard Oil of Ohio license for power saving units in electronics, and this application was considered as the starting point for commercial based electrochemical capacitors (ECs) [7]. In 1970, after few modifications, the first electrochemical capacitor patented by Donald L. Boos as 'Electrolytic Capacitor Having Carbon Paste Electrodes' [9].

Presently, the fast growth in mobile phones, electronic devices and electric vehicles created a need for innovative electrochemical energy storage devices with high

© The Authors 2020
V. V. Jadhav et al., *Bismuth-Ferrite-Based Electrochemical Supercapacitors*, SpringerBriefs in Materials, https://doi.org/10.1007/978-3-030-16718-9_2

Fig. 2.1 Block diagram of
capacitor patented by general
electric (adapted from [6])

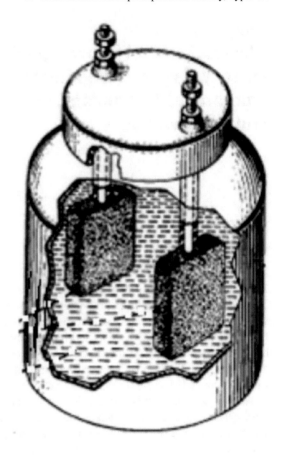

power capabilities. Also, the increase in demand for supercapacitors in energy harvesting applications (wind and solar energy) and use of supercapacitor in aircrafts and train influence the global supercapacitor market attracted the attention of the researchers.

In the early 1990s, the United States Department of Energy (DOE) created international awareness of the potential of the supercapacitor and battery research and strongly supported funding for battery and supercapacitor research. Then in 1992, Maxwell implemented the term ultracapacitor from PRI and called 'Boots Caps' to feature their use for power applications which results into invention and development of innovative cost-effective electrode material and electrolyte which can boot the electrochemical performance. Maxwell's core business is ultracapacitors with high power density energy storage devices with a wide temperature range that can rapidly charge and discharge. In 2019, Tesla announced a plan to acquire Maxwell technologies to established ultracapacitor and storage material firm for $218 million in an all-stock deal. Today, several companies such as TESLA, Maxwell Technologies, Panasonic, Nesscap, EPCOS, NEC, ELNA, and TOKIN invest heavily in electrochemical capacitor development.

The worldwide supercapacitor market was valued at USD 685 million in 2018, and it is predictable to reach USD 2187 million by 2024, registering a CAGR of 21.8%, during the period of 2019–2024. With high temperature stability, power, and energy density, supercapacitors replace the traditional car batteries. In the near future, supercapacitor could be alternative to lithium-ion battery because of safety issues. Moreover, supercapacitors are more versatile and stable than normal batteries which have great demand in mobile device, portable media players, GPS, laptops, etc.

2.2 Types

A supercapacitor, also known as ultracapacitors or electrochemical capacitor, is an energy storage device, which can act as a gap bridging function between batteries and conventional capacitors [10]. Depending on the charge storage mechanism and research and development trends, electrochemical capacitors are classified into three types, namely;

(a) electrical double-layer capacitors,
(b) pseudocapacitors (PCs), and
(c) hybrid capacitors as shown in Fig. 2.2.

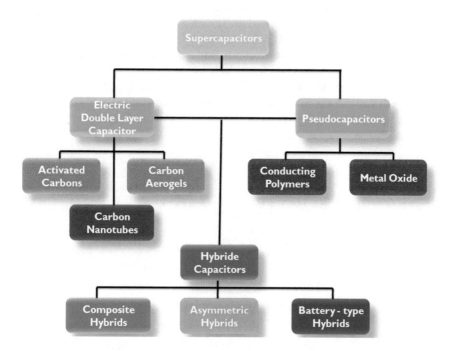

Fig. 2.2 Types of a supercapacitor

The classification of electrochemical capacitors is based on its charge storage mechanism such as are faradic, non-faradic, and a combination of both. The redox reaction is the faradic process in which the transfer of charge takes between electrode and electrolyte and the capacitance comes from electrostatic charge gathered at electrode/electrolyte interface. Thus, faradic process is sturdily depending on surface area of the electrode material. However, in non-faradic process charges are physically spread over electrode surface and is completely isolated from chemical mechanism as it does not involve any making/breaking of chemical bonds.

2.2.1   *Electrical-Double Layer Capacitor (EDLC)*

These capacitors depend on carbon-based structures exploiting non-faradic electrostatic charging of the electrical double-layer formed at the electrode–electrolyte interface and are hence termed as electrical double-layer capacitors. EDLC involves only physical adsorption of ions without any chemical reactions. The EDLC is associated with an electrode-potential-dependent accumulation of electrostatic charge at the interface. The mechanism of surface electrode charge generation includes surface separation as well as ion adsorption from both the electrolyte and crystal lattice defects [11]. These processes work entirely on the electrostatic accumulation of surface charge. The EDLC generally appears between electrode material and electrolyte where the electric charges are gathered, and to achieve the electroneutrality, counterbalancing charges are assembled upon the electrolyte side. During charging process, cations move toward the negative electrode, while anions move toward the positive electrode. During discharge, said process gets reversed.

If the two electrode surfaces can be expressed as E_{S1} and E_{S2}, an anion as A^-, a cation as C^+, and the electrode/electrolyte interface, the electrochemical processes for charging and discharging can be expressed as (2.1)–(2.6) [12, 13].

On one electrode (a positive one),

$$E_{S1} + A^- \xrightarrow{\text{Charging}} E_{S1}^+ //A^- + e^- \tag{2.1}$$

$$E_{S1}^+ //A^- + e^- \xrightarrow{\text{discharging}} E_{S1} + A^- \tag{2.2}$$

On the other electrode (say, a negative one),

$$E_{S2} + C^+ + e^- \xrightarrow{\text{Charging}} E_{S2-} //C^+ \tag{2.3}$$

$$E_{S2-} //C^+ \xrightarrow{\text{Discharging}} E_{S2} + C^+ + e^- \tag{2.4}$$

And the overall charging and discharging process can be expressed as (2.5) and 2.6),

$$E_{S1} + E_{S2} + A^- + C^+ \xrightarrow{\text{Charging}} E_{S1}^+ // A^- + E_{S2}^- // C^+ \qquad (2.5)$$

$$E_{S1}^+ // A^- + E_{S2}^- // C^+ \xrightarrow{\text{Discharging}} E_{S1} + E_{S2} + A^- + C^+ \qquad (2.6)$$

In this type of ESs, there is no charge transfer across the electrode/electrolyte interface, and also no net ion exchange takes place between the electrode and the electrolyte. This denotes that the electrolyte concentration remains constant during the charging and discharging processes. In this way, energy can be stored in the double-layer interface.

2.2.2 Pseudocapacitor

The operation of pseudocapacitors (PCs) is different than EDLCs. When a potential is applied to a PCs, fast and reversible faradic reactions (redox reactions) take place on the electrode materials and involve the passage of charge across the double layer, similar to a process taking place in batteries, resulting in faradic current passing through the supercapacitors cell. The charge transfer that takes place in these reactions is voltage-dependent, so a capacitive phenomenon occurs [14–16]. Materials undergoing such redox reactions include conducting polymers and several metal oxides, including RuO_2, MnO_2, and Co_3O_4 [17–19]. Three types of faradic processes found at PC electrodes—reversible adsorption, redox reactions of transition metal oxides (e.g., RuO_2), and reversible electrochemical in conductive polymer-based electrodes [20]. It has been demonstrated that these faradic electrochemical processes not only extend the working voltage but also increase the specific capacitance (SC) of the supercapacitors [21]. Since the electrochemical processes are taking place both surface and in the bulk near the surface of the solid electrode, a PC exhibits far higher capacitance and energy density than an EDLC. Conway et al. studied the capacitance of PCs which is 10–100 times higher than the electrostatic capacitance of an EDLC [22]. Nevertheless, PCs usually suffer from relatively lower power density than EDLC because faradic processes are normally slower than non-faradic processes [23]. Moreover, because redox reactions occur at the electrode, pseudocapacitors often lack stability during cycling, similar to batteries.

2.2.3 Hybrid Electrochemical Capacitor

A combination of faradaic and non-faradaic mechanisms would generate supercapacitors that exhibit high capacitance for pulse power as well as sustained energy. An asymmetric electrode arrangement where one electrode consists of electrostatic carbon material, while the other consists of faradaic capacitance material. These supercapacitors are mentioned to as hybrid supercapacitors. In this kind of hybrid supercapacitor, both electrical double-layer capacitance and faradic capacitance mechanisms occur simultaneously, but one of them plays a dominating role. In both cases, a large surface area, high conductivity, and appropriate pore-size distribution are essential criteria of the electrode materials to achieve high capacitance. In recent years, several types of hybrid supercapacitors or hybrid electrochemical capacitors with an asymmetrical configuration have attracted noteworthy attention [24, 25]. Until now, most of these capacitors developed as the cathode material. The PC electrodes accumulate charge through a faradic electrochemical process, which can not only increase the SC of the capacitor but also extend the working voltage. In an advanced hybrid supercapacitor, the potential range at the cathode is extended to the whole potential window of activated carbon, specifically from 1.5 to 4.5 V versus Li/Li^+, which is wider than the conventional EC, where the potential varies from 0.8 to 2.7 V versus Li/Li^+. Briefly, the energy density of SC devices is higher than for EDLCs.

2.3 Operation Mechanisms

Electrochemical capacitors work on a principle similar to those of conventional electrostatic capacitors. It is, therefore, useful to undertake a brief review of the electrostatic capacitor operation. Conventional capacitor stores energy in the form of electrical charge, and a typical device consists of two conducting materials separated by a dielectric as shown in Fig. 2.3.

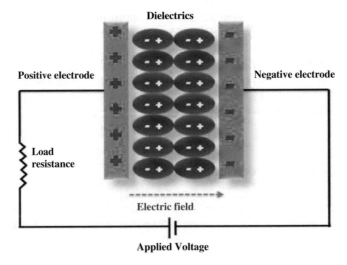

Fig. 2.3 Schematic of conventional capacitor

When an electric potential is applied across the conductor electrons begin to flow and charge accumulates on each conductor and after removing potential, the conducting plates remain charged till brought into contact again where energy get discharged. The amount of charge that can be stored concerning the strength of the applied potential is known as the capacitance and is a measure of a capacitor's energy storage capability.

Equations (2.7) and (2.8) apply to an electrostatic capacitor.

$$C = \frac{Q}{V} = \varepsilon \frac{A}{D} \tag{2.7}$$

$$U = \frac{1}{2}CV^2 = \frac{1}{2}QV \tag{2.8}$$

where, C is the capacitance in Farads, Q is charge in Coulomb, V is electric potential in volts, ε is the dielectric constant of the dielectric, A is conductor surface area, d is dielectric thickness, and U is the potential energy. Therefore, the factors affecting on the capacitance are:

- plate area (common to the two electrodes),
- the separation distance between the electrodes, and
- properties of the dielectric (inductor) used.

Through reversible adsorption of ions, the ESs store the electric energy which forms an electric double layer at electrode/electrolyte interface. ES consists of two electrodes, an electrolyte, and a separator that electrically isolates the two electrodes (Fig. 2.4). The most important component in the ESs is the electrode materials which can be metal oxides/chalcogenides, halides, MXene, 2D and 3D electrode, layered double hydroxides, etc., of high surface area and high porosity.

Figure 2.4 shows that the charges are stored and separated at the interface between the conductive solid particles (such as carbon particles or metal oxide particles) and the electrolyte. This interface can be treated as a capacitor with an electrical double-layer capacitance, which can be expressed as (2.9)

$$C = \frac{A\varepsilon}{4\pi d} \tag{2.9}$$

where, A is the area of the active electrode surface, ε is the medium (electrolyte) dielectric constant equal to 1 for a vacuum and larger than 1 for all other materials, and d is the actual thickness of the electrical double layer.

2.4 Designing Processes and Working Mechanisms

The EDLC consists of two electrodes that supposed to be in an electrolyte with a separator between them. The electrode consists of a current collector in contact with the active electrode material. The interfacial separation of electronic and ionic

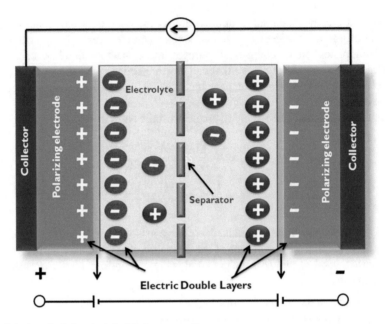

Fig. 2.4 A typical electrical double-layer capacitor

charges is mainly responsible for energy storage in EDLCs. Moreover, in EDLCs, the charge separation distance (d) can be reduced to the Helmholtz double-layer (d_D) thickness, which is defined as half the diameter of the adsorbed solvated ions at the electrode/solution interface [26]. This assembly is placed in a tight casing to avoid liquid leakage and gas. While manufacturing a supercapacitor, it is necessary to distinguish different steps:

- manufacturing of the electrode,
- winding or stacking,
- filling with an electrolyte,
- testing, and
- welding and sealing.

2.4.1 Manufacturing of the Electrodes

The most important component of the supercapacitor is the electrode material which acts as a current collector. The electrode determines the ES performance in terms of self-discharge, life expectancy, capacity, resistance, and so on. Therefore, electrode fabrication including an active material coating on current collector process is the most important step. As a result, strictly controlling the preparation process is necessary for achieving both high performance and stability. Generally, aluminum, nickel, copper, and stainless steel are used as a current collector during fabrication. But the most useful material for current collectors is aluminum. The aluminum foil is used as the current collector because of its high current-carrying capability, chemical stability, and low cost. Major industrially used current collectors have a special surface state in order to increase the mechanical adhesion between the active electrode and current collector.

2.4.2 Winding or Stacking

During the winding of electrodes, binders, active materials, and conductive additives are majorly mixed to obtain homogeneous slurry with the desired density. Then, this slurry is coated onto an etched aluminum foil, dried, and rolled to achieve a uniform electrode coating layer.

The prepared electrodes taking into glove box with ultralow moisture followed by a separator layer inserted between them are wounded around a central mandrel into the desired shape. The main advantage of the coating (winding) is to control the thickness of the electrode materials to achieve a good volume density of power and energy. The most popular condensed liquid binder in the supercapacitor technology is polyvinylidene fluoride and polytetrafluoroethylene because of its high electrochemical inertia and ability to keep a process in aqueous media [27].

2.4.3 Filling with an Electrolyte

After completion of the winding process, the electrolyte is filled into the system. The electrolyte-filling process generally requires special care. The amount of electrolyte in the cell is critical because excess electrolyte can lead to excessive gassing and leakage in the process. Generally, the process is time-consuming and is relatively labor-intensive. This is because of the nanopore's level of internal porosity of active materials, making the wetting process slow and need to be soaked around 24 h.

2.4.4 Testing

After the completion of electrolyte filling, the cell is subjected to a cycling test (for 2–5 formation cycles). During formation, the abnormal cells are removed away. The manufacturing will be formed with evacuation with final welding and sealing. In order to match different applications, ESs should be connected in series or in parallel to obtain the desired output voltage or energy capacity.

2.4.5 Welding and Sealing

This is the last step in which a supercapacitor is manufactured. The rating and applications, the parameters which can vary commercial products and research devices, are strongly affecting factors on the design, assembly, and packaging of the supercapacitor. Figure 2.5 shows details of a basic single supercapacitor cell. Commercially available supercapacitors often prefer stacked cylindrical or coin form. Each manufacturer has its own propriety packaging systems. Cylindrical, typically called 'jelly rolls' because of their shape require the formation of the electrode layer film by rolling or spraying a carbon material on both sides of a separator. An outer separator is applied to ensure the layers are electrically isolated from each other.

While preparing coin cell assembly, the cells get pressed and sealed to achieve good electrical contacts and prevent electrolyte leakage, respectively. The coin cell, with simple construction, can be prepared in any ordinary laboratories. A direct laser welding between the cover and current collector without any intermediate parts has been developed by Batscap to limit the equivalent series resistance. Maxwell Technologies and Nesscap have produced the same design using wounding welded on an intermediate aluminum stamped part on the cover. The main manufacturers are Nesscap (South Korea), Batscap (France), Maxwell Technologies (USA), and LS Mtron (South Korea). In 2019, Tesla paid over $200 million for ultracapacitor manufacturer to brought Maxwell Technologies. In electric vehicles, supercapacitors are predominantly used for regenerative braking.

Fig. 2.5 Basic (single) supercapacitor cell

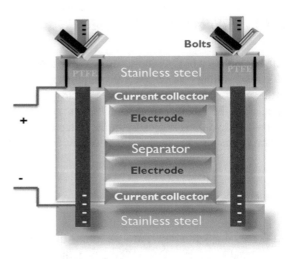

2.5 Supercapacitor Electrode Materials

Electrode materials for supercapacitors are majorly carbonous, metal oxides, and conducting polymers, which generate three categories of capacitors—EDLC-based carbon materials, PC based on either transition metal oxides or conducting polymers, and hybrid capacitors (HCs). Composite HCs employ a double-layer-type electrode against a PC electrode, such as carbon-MnO_2 [28, 29] and Zn_2SnO_4/MnO_2 [30]. Advanced asymmetric $MnCo_2O_{4.5}/Ni(OH)_2$ electrodes of battery–supercapacitor hybrid devices [31].

The electrode materials need to have following requirements:

- high specific surface area,
- suitable surface functional groups to enhance the capacitance by additional faradic redox reaction with improved surface wettability,
- low internal electric resistance, which leads to fast charging–discharging and low ohmic resistance,
- low price,
- low volume and weight,
- environmentally friendly materials, and
- thinner electrodes and current collectors.

2.5.1 Carbonaceous Materials

Carbonaceous materials are widely utilized for electrode materials in supercapacitors due to their easy accessibility, good processing ability, large surface area, porosity, low electrical resistivity, robust surface chemical environment, physicochemical stability, and low cost. The energy storage capability of carbon materials is owing to the electric charges stored across the electrode/electrolyte interface. The average capacitance of the carbon electrode is 50–200 F g^{-1}, 30–100 F g^{-1}, and 20–70 F g^{-1} in aqueous, organic, and ionic electrolytes, respectively.

The carbon materials used include activated carbons, carbon nanotubes (CNTs), graphene, carbon nanofibers (CNFs), carbon aerogels, ordered mesoporous carbons, and hierarchical porous carbons. Activated carbons can often achieve maximum capacitances about 100–200 F g^{-1} in aqueous electrolyte and 50–150 F g^{-1} in the organic medium [32]. Highly porous activated carbon prepared by Maloney et al. [33] showed 394 F g^{-1} at a current density of 1 A g^{-1} with excellent cycling stability and 94% retention in 6 M KOH after 5000 cycle derived from willow wood. Hu et al. [34] synthesized N-doped porous carbons by a one-step sodium amide activation process using lotus leaf as the raw material exhibit 266 F/g and excellent stability in cycling test retention of 97%.

Activated carbon (AC) electrodes are commonly used in EDLCs because they are economical and possess a high surface area. ACs have a complex porous structure composed of micropores (<2 nm), mesopores (2–50 nm), and macropores (>50 nm). For AC-based devices, it is found that larger pore sizes correlate with higher power density and smaller pore sizes correlate with the higher energy density and hence, the optimization of the pore size became essential [35]. The various porous structures can be prepared by both physical and chemical [36–38]. The CNT is 1D, and graphene and MXene are 2D carbon materials which have vigorously attempted as electrode materials in energy conversion/storage systems as the positive and negative electrode, owing to their excellent integrated electronic, mechanical, and chemical properties [39–42]. Graphene, a new class of carbon material, exhibits good electrical conductivity ranging from 200 to 3000 S cm^{-1}. Chemical exfoliation of graphite enables mass production of chemically modified graphene with a large quantity of exposed active sites anchored with oxygen functional groups [43]. The robust surface chemical properties of graphene materials make them feasible as supercapacitor electrode materials [44–46]. These fascinating features make it quite promising as a soft electrode for flexible supercapacitors.

2.5.2 Conducting Polymers

Conducting polymers are another kind of electrode material that are capable of delivering faradic charge. The applicable conducting polymers for PCs mainly include polyaniline (PANI), poly (3,4-ethylenedioxythiophene), and polypyrrole. Similar to oxide materials, the problems of conducting polymers are relatively low electrical conductivity, low stability, and high cost in comparison with carbon materials. PANI has been widely studied as positive electrode material in literature due to its ease of synthesis and high capacitance (3407 F g^{-1}) [47, 48]. It has many desirable properties for use in a supercapacitor device including high electroactivity, a high doping level, excellent stability, and specific capacitance (600 Fg^{-1} in an acidic medium) [49]. In addition, it has good environmental stability and controllable electrical conductivity and can be easily processed [50, 51]. A major disadvantage of PANI is that it requires a proton to be properly charged and discharged; therefore, aprotic solvent, an acidic solution, or a protic ionic liquid is required [52]. Polypyrrole has received great attention for its high capacitance, easy fabrication process, better chemical and thermal stability than most conducting polymer for energy storage application such as supercapacitor and battery research [53–57].

2.5.3 Metal Oxides

The metal oxides are well-known materials that are extensively used not only in other storage devices like magnetic and optical but also in electrochemical too. The surface of metal oxide is a key factor for effective interaction with target molecules. Engineering the surface of metal oxide at nanoscale increases the surface area which improves the electrochemical properties of the active electrode material. Hence, the metal oxides came forward as an emerging material with exclusive properties. The most common and promising metal oxide used for supercapacitor is the transition metal oxide such as RuO_x, NiO, MnO_x, FeO_x because they exhibit PCs and EDLC [58–72]. The most attractive advantage of metal oxides to carbon materials is their much larger capacitance, due to multielectron transfer during fast faradic reactions. RuO_2 shows excellent supercapacitive properties, such as high specific capacitance and very long cycling life, but unfortunately, the high cost, the poor cycle life of RuO_2 limits its commercial potential. The cheaper and abundant metal oxides suffer from poor electrical conductivity and poor cycling stability problems, which prevents the high electrochemical performance of these materials. In addition to the above metal compounds, Ir, Co, Mo, Ti, V, Sn, Fe, and other metal oxides can also be used as electrode materials for supercapacitors. Nitrides (titanium nitride, iron nitride, etc.) are good PCs as well [73]. Very appealing electrochemical characteristics and commercial applicability of RuO_2 are so far hindered by its cost. Ruthenium containing ternary oxides and composites helps to reduce cost and increase the utilization of RuO_2, although these approaches undermine the specific capacitance to some extent [74–77].

The manganese oxide has the tunnel and layered crystalline structures that constitutes a large family of porous materials, from ultramicropores to mesoporous. Depending on the linkage of the MnO_6 octahedral structural units, MnO_2 can exist in diverse crystalline phases that exhibit different cation exchange, adsorptive and electrochemical responses [78–81]. MnO_2 exhibits PC behavior owing to the surface adsorption and insertion/extraction of metal alkali cations inside the open crystalline network. Due to the low diffusion coefficient of protons and alkali cations to the interior of MnO_2, the specific capacitance of conventionally prepared MnO_2 in the presence of binder and conductive additives is often inhibited to below 200 Fg^{-1}, while binder-free nanostructured MnO_2 with reduced electronic and charge diffusion distances is expected to promote charge-transfer reactions [82–86]. Hence, MnO_2 is a technologically important and promising material for energy storage and catalysis application [87]. The other metal oxides that received considerable attention as an alternative electrode that include Bi_2O_3 [88], $BiFeO_3$ [89], Co_3O_4 [90], Fe_3O_4 [91], V_2O_5 [92], and NiO [93]. The other types of metal oxides are $MnCo_2O_4$ [2], $CuCo_2O_4$ [94], $NiCo_2O_4$ [95], and metal hydroxides like α-$Co(OH)_2$/α-$Ni(OH)_2$ [96], $Mn(OH)_2$ [97], $Cu(OH)_2$ [98], and $Ni(OH)_2$ [99]. Despite huge efforts made on oxide materials, the research community still continues to explore new electrode materials that are cheap, abundant, and possess good electrochemical performance.

2.5.4 Hybrid/Composite Materials

In recent years, several types of HCs with an asymmetrical configuration have attracted significant attention [100]. The PC electrodes accumulate charge through the faradaic electrochemical process, which can not only increase the specific capacitance (SC) of the capacitor but also extend the working voltage. The supercapacitor–battery hybrid device has potential applications in energy storage and can be a remedy for low-energy supercapacitors and low-power batteries [4]. Also, MXene-based hybrid supercapacitor shows exceptional flexibility and integration for high-performance capacitance and voltage output [101]. These results provide the possibility of designing high-energy-density hybrid asymmetric supercapacitors for practical device applications.

2.6 Electrolytes

The capacitance of the EDLC strongly depends on the electrolyte. The ability to store charge is dependent on the accessibility of the ions to the porous surface area; thus, ion size and pore size must be optimal. So, both electrode and electrolyte must be chosen together. Poor electrolyte stability at different cell operating temperatures and poor chemical stability at high rates can further increase resistances within an ES and thus, reducing cycle life. The electrolyte with good chemical and electrochemical stabilities allows larger potential windows. To ensure the safe operation of ESs, electrolyte materials should have low volatility, low flammability, and low corrosion potential. Each solvent exhibits varying levels of ionic conductivity, voltage stability, size, and reaction concerns that must be considered when choosing an electrolyte. The electrolytes used in an ES are of four types; (1) aqueous, (2) organic, (3) ionic liquids, and (d) solid-state polymer electrolyte.

2.6.1 Aqueous Electrolytes

The aqueous electrolytes such as KOH, NaCl, KCl, K_2SO_4, H_2SO_4, NaOH, Na_2SO_4, and NH_4Cl can provide a higher ionic concentration and lower series resistance [102, 103]. ESs containing aqueous electrolyte display higher capacitance and higher power density than those with organic electrolytes due to higher ionic concentration and smaller ionic radius. The disadvantage of aqueous electrolytes is their small voltage window of about 1.2 V, much lower than those of organic electrolytes.

2.6.2 Organic Electrolytes

Ethylene carbonate, acetonitrile, tetrahydrofuran, and propylene carbonate are generally used as an organic electrolyte to enhance supercapacitor performance [104]. Among organic electrolytes, acetonitrile and propylene carbonate are the most commonly used solvents. The acetonitrile is good electrolyte but is environmentally toxic.

Propylene carbonate-based electrolytes are friendly to the environment and can offer a wide electrochemical voltage window, a wide range of operating temperatures, as well as good conductivity. Besides, organic salts such as tetraethylammonium tetrafluoroborate, tetraethyl phosphonium tetrafluoroborate, and triethylmethylammonium tetrafluoroborate have also been used in ESs as electrolytes. However, one issue which would keep in mind is that the water content in organic electrolytes should be below 3–5 ppm to avoid the voltage drop.

2.6.3   Ionic Liquids

Ionic liquids (ILs) are more appropriate electrolyte which can be used in ESs. A pinch of salt may be melted, namely 'liquefied,' by providing heat to the system to counterbalance the salt lattice energy. Such systems are referred as molten salts or ILs which exist in liquid form at the particular temperatures. ILs exist as viscous molten salts at ambient temperatures, allowing heavy concentrations in solvents or removal of solvents altogether. Low vapor pressure, low flammability, and low toxicity keep health risks low. High chemical stability of ILs allows operation at a higher voltage window (5 V). The main ILs studied for ES applications include imidazolium, pyrrolidinium, as well as asymmetric, aliphatic quaternary ammonium salts with anions such as tetrafluoroborate, trifluoromethanesulfonate, bis(trifluoromethanesulfonyl)imide, bis(fluorosulfonyl) imide, or hexafluorophosphate [105].

The most studied ionic liquids are the 1-Ethyl-3-methylimidazolium tetrafluoroborate. The major drawback of ionic liquids is their low conductivity at room-temperature in aqueous and acetonitrile-based systems [106]. IL-based electrolytes demonstrate high thermal stability that creates an opportunity for operation in high-temperature environments. At high temperatures, the low conductivity limits IL performance which can overcome with ion mobility for better device power and response time [106]. However, a high heat reduces the potential window for ion stability, thereby impacts the power and energy density. Another way to overcome the low conductivity of ILs is to balance their high potential window on increasing conductivity with organic electrolytes such as propylene carbonate and acetonitrile [107].

2.6.4   Solid-State Polymer Electrolytes

Solid-state polymer supercapacitors have vast potential for future development due to their environmentally friendly, compact, safe, and facile designing. It is expected that the solid-state supercapacitor system may provide a scalable strategy towards powering future wearable electronic devices. Gel and solid polymer electrolytes aim to combine the function of the electrolyte and separator into a single component to reduce the number of parts in an ES and enhance the potential window over the

higher stability. A gel electrolyte incorporates a liquid electrolyte into a microporous polymer matrix that holds in the liquid electrolyte through capillary forces, creating a solid polymer film. Gel electrolytes allow the incorporation of aqueous, organic, and ionic liquids, depending on the requirements of the ES. Separators are used in conjunction with the electrolyte to prevent short-circuiting between the electrodes. In order to combine the two structures, electrolyte needs to be trapped within the polymer matrix during polymerization. The result is a solid, thin, flexible electrolyte. Gel polymers offer slightly lower conductance than liquid electrolytes, but they provide structural improvement that improves the efficiency of ion transport mechanisms and cycle life [108–110].

2.7 Separator

Recently, ES has limited operating temperature mainly because of the thermal stability of the separator which hinders the performance at high temperature. The electrochemical stability, high porosity, chemical and thermal inertia are the essential properties of the separators in ES.

In addition, because a separator is a non-reactive material in a supercapacitor, the separator thickness must be as thin as possible and the cost must be low, as expected. The separator should prevent the occurrence of electrical contact between the two electrodes, but it must be ion-permeable and allow ionic charge transfer to take place. Polymer or paper separators are used with organic electrolytes and ceramic or glass fiber separators are often used with aqueous electrolytes. For best EDLCs performance, the separator should have high electrical resistance, a high ionic conductance, and a low thickness [111]. Theoretically, supercapacitor electrodes can work without a separator if the distance between electrodes is optimum [112]. Practically, it is quite difficult to maintain such devices without any short circuits. Chen et al. reported the fabrication of high-temperature ceramic separator can withstand over 350 °C without any shrinkage [113].

2.8 The Energy Storage Mechanisms

2.8.1 EDLCs

Conventional capacitors store little energy due to the limited charge storage areas and geometric constrains of the separation distance between the two charged plates. However, supercapacitors based on the EDLCs mechanism can store much more energy because of the large interfacial area and the atomic range of charge separation distances. As schematically illustrated in Fig. 2.6a the concept of the EDLCs was first described and modeled by Von Helmholtz in the nineteenth century [114], when he

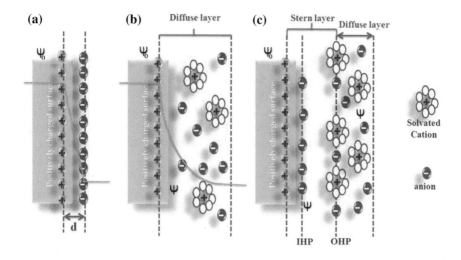

Fig. 2.6 Models of the electrical double layer at a positively charged surface; **a** the Helmholtz model, **b** the Gouy–Chapman model, and **c** the Stern model, showing the inner Helmholtz plane and outer Helmholtz plane

investigated the distribution of opposite charges at the interface of colloidal particles. When two layers of opposite charges formed this interface separated by an atomic distance, then such model is referred as the Helmholtz-double-layer model which was further modified by Gouy and Chapman [115, 116]. This simple Helmholtz EDLCs model was further modified by Gouy and Chapman [115, 116]. On consideration of the continuous distribution of electrolyte ions in the electrolyte solution driven by thermal motion is referred to as the diffuse layer as shown in Fig. 2.6b. However, the Gouy–Chapman model leads to an overestimation of the EDL capacitance. The capacitance of two separated arrays of charges increases inversely with their distance of separation; hence, a very large capacitance can obtain in the case of point charge ions close to the electrode surface. Later and Stern [117], combined the Helmholtz model with the Gouy–Chapman model to explicitly recognize two regions of ion distribution. The inner region is called the compact layer or Stern layer and the diffuse layer as shown in Fig. 2.6c. In the compact layer, ions are strongly adsorbed by the electrode, thus the name of the compact layer. In addition, the compact layer consists of specifically adsorbed ions (anions irrespective of the charged nature of the electrode) and non-specifically adsorbed counter ions. The inner Helmholtz plane (IHP) and outer Helmholtz plane (OHP) are used to distinguishing the two types of adsorbed ions. The diffuse layer region is what the Gouy–Chapman model defines. The capacitance in the EDL (C_{dl}) can be treated as a combination of the capacitances from two regions, the Stern type of compact double-layer capacitance (C_H) and the diffusion region capacitance (C_{diff}). Thus, C_{dl} can be expressed by the following equation:

$$\frac{1}{C_{dl}} = \frac{1}{C_H} = \frac{1}{C_{diff}}$$ (2.10)

The factors that determine the EDL behavior at a planar electrode surface include the electrical field across the electrode, the types of electrolyte ions, the solvent in which the electrolyte ions are dissolved, and the chemical affinity between the adsorbed ions and the electrode surface. The main difference between PC and EDL capacitance lies in the fact that PC is faradic in origin, involving fast and reversible redox reactions between the electrolyte and electro-active species on the electrode surface.

2.8.2 PCs

As discussed earlier, the origin of PCs is a reversible faradic reaction occurring at the electrode. The charge transfer that takes place in these reactions is voltage-dependent, so a capacitive phenomenon occurs. The voltage-dependent charge transfer reactions are of two types and are as follows:

2.8.2.1 Redox Reactions

In a redox reaction consists of oxidant, ox, and reluctant, red, as

$$ox + ze^- \leftrightarrow red$$ (2.11)

The potential, E, is given by the Nernst equation as shown in (2.12)

$$E = E^O + \frac{RT}{zF} \ln \frac{k}{1-k}$$ (2.12)

where, E^o is the standard potential, R is the gas constant, T is the absolute temperature, F is the Faraday constant, and $k = [ox]/([ox] + [red])$, where square bracket presents concentration. The amount of charge q (given by the product zF), therefore, is a function of the potential E. Differentiation of (2.12) thus produces a pseudocapacitive relation [118, 119].

2.8.2.2 Adsorption of Ions

The deposition of ions to form a monolayer on the electrode substrate is a reversible process that results in a Faradic charge transfer and hence gives rise to pseudoca-pacitance in a similar manner to that demonstrated in redox reactions. The most commonly known active species are ruthenium oxide [120], manganese oxide [121], vanadium nitride [122], electrically conducting polymers such as PANI [123] and oxygen or nitrogen-containing surface functional groups [124].

2.9 Advantages

2.9.1 High Power Density

It is clear that ESs display a much higher power delivery ($1–10\,kW\,kg^{-1}$ as compared to lithium-ion batteries ($150\,W\,kg^{-1}$). An ES stores the electrical charges at electrode surface instead of within entire electrode, the charge–discharge of the electrode material is not limited by ionic conduction into the electrode bulk which improves the charging and discharging rates effectively. These rapid rates lead to a high-power density in ES. For example, an ES can be fully charged or discharged in few seconds (30 s), and the energy can be taken out from it very rapidly, within 0.1 s [125]. For batteries, the charging time is normally on the scale of hours.

2.9.2 Long Life

ESs need not require any maintenance during their lifetimes and can withstand a huge number of charge–discharge cycles, up to 1,000,000. Moreover, ESs can run deeply at high rates for 500,000–1,000,000 cycles with only small changes in their characteristics. Such longevity is impossible for batteries even if the depth of discharge is as small as 10–20% of the overall energy. The life expectancy for ES is estimated to be up to 30 years, which is much longer than for lithium-ion batteries (1000–10,000 cycles and a life expectancy of only 5–10 years). Even in ESs, although fast redox reactions are involved during charging and recharging, their life expectancy is also much longer than that of batteries [126].

2.9.3 Long Shelf Life

Another advantage of supercapacitors is their long life. In contrast, ESs maintain high capacitance and thus are capable of being recharged to their original condition, although self-discharge over a while can lead to a lower voltage.

2.9.4   High Efficiency

ES is reversible with respect to charging and discharging throughout its complete operating range of voltage, and the energy lost to heat during each cycle is relatively small and readily removed.

2.9.5   Wide Operating Temperatures Range

ES can work excellently in any temperatures. The typical operating temperature for ES ranges from -40 to $200\ ^{\circ}C$ [127] which is useful in military applications during war.

2.9.6   Environmental Friendliness

In general, ES designs are free from hazardous or toxic materials, and their waste materials are easily disposed.

2.9.7   Safety

In normal circumstances, ESs are much safer and faster than batteries, in particular, lithium-ion batteries. Compared to other energy storage systems such as batteries and fuel cells, ESs are superior in the areas of life expectancy without any noticeable performance degradation after the long-term operation, reversibility, power density, shelf life, efficiency, operating temperature at the range, environmentally friendliness, and safety.

2.10   Applications

With the increasing demand for energy coupled with power shortages and high prices in the globalized world, there has been a stimulating drive to research advanced energy storage and management devices. Supercapacitors have attracted particular attention in light of their excellent power density, fast charging rate, and extended cycle life. The major applications of electrochemical capacitors appear to be in high-pulse power and short-term power hold. A few applications of electrochemical capacitors are discussed below in detail.

2.10.1 Memory Backups

Commercially many appliances are integrated with components with memory where a small disruption in power supply can cause the loss of stored information. In such applications, the capacitor acts as a backup supply for short periods. Batteries can be a good option for such applications but because of short lifetime, batteries need to be replaced regularly. This shortcoming of batteries can be solved by ESs which has longer lifetime.

2.10.2 Battery Improvement

Nowadays, the batteries are extensively used in portable power appliances, for instance laptops and mobile phones. Batteries are found equivalent to an electrochemical capacitor which can effectively be the substitute for such applications.

2.10.3 Portable Power Supplies

These days, the available power supply consuming comparatively long charging time which need to be reduced by means of electrochemical capacitors which can be the effective alternative for the standalone power sources

2.10.4 Renewable Energy Applications

Owing to damaging effect on batteries solar photovoltaic applications, it needs to be replaced after few years. The cost-effective electrochemical supercapacitors, which have the small recharge time with long discharge time, can became the promising alternative with long life spam which is well equivalent to photovoltaic panels.

2.10.5 Military and Aerospace Applications

With the development of technology, the military market segment has demonstrated increasing needs for ES devices to address a variety of power challenges based on their unique characteristics. They provide quick power delivery wide temperatures operation range, and the ability to handle up to a million cycles. In military applications, the most frequent ESs use is to provide a backup power for electronics in military vehicles, fire control systems in tanks and armored vehicles, airbag deployments and helicopters black boxes, and memory devices for handheld radios. The ES-based modules are also appropriate in peak power applications to facilitate reliable communication transmission on land-based vehicles, active suspension for vehicle stabilization, cold engine starts, and bus voltage hold-up during peak currents.

2.11 Disadvantages

The supercapacitor has many advantages over batteries and fuel cells and also a few disadvantages.

2.11.1 Low Energy Density

Supercapacitors suffer from limited energy density (about 5 Wh kg^{-1}) when compared with batteries (>50 Wh kg^{-1}), as shown in Fig. 1.2 (region plot, Chap. 1). Commercially available ES can provide energy densities of only 3–4 Wh kg^{-1}. If a large energy capacity is required for an application, a heavy supercapacitor must be constructed. Low energy density is the major disadvantage for supercapacitor applications in the short and medium terms.

2.11.2 High Cost

The costs of raw materials and manufacturing processes are major concerns of ES technology. At present, for practical purposes, carbon and RuO_2 are being commonly used. However, carbon materials with a high surface area are not expensive. Few rare metal oxides such as RuO_2 are very expansive. In addition, the separator and the electrolyte can also boost the expense. For example, if ESs uses organic electrolytes, their cost is far from negligible.

2.11.3 High Self-discharging Rate

ESs have a low duration and a high self-discharging rate of 10–40% per day which has been considered as a foremost impediment to practical use.

2.11.4 Industrial Standards for Commercialization

Currently, carbon-based ES with the capacitance of 50–5000 F has become commercially available. The electrolyte used in such kind of ES is acetonitrile which can give a cell voltage of about 2.7 V with a typical specific energy of 4 Wh kg^{-1}. This performance can be achieved on charging the device at 400 W kg^{-1} from 100 to 50% of the rated voltage. Although this kind of ES is commercially available, it is necessary to establish some general industrial standards such as performance, electrode structure, electrode layer thickness, and porosity. However, due to the variety of applications as well as limited commercial products, we are still not able to search out generally available industrial standards for ES at this moment. Therefore, it is necessary to put some effort into ES standard establishment for different applications.

2.12 Conclusions

This chapter has reviewed history, different types, designing processes, operation mechanisms, advantages and disadvantages of electrochemical supercapacitors. It explains the essential requirements of supercapacitors. Also, recent progress in the charge storage mechanisms, active materials, electrolytes used in electric double-layered capacitor, pseudocapacitor, and hybrid capacitors are explained in details. Supercapacitor can be used for different applications such as uninterruptible power supplies, mobile phones, hybrid vehicles, military warheads, solar cells. Despite of their practical advantages, the low energy density and high cost avoid them from a total replacement of batteries. Those main challenges must be overcome without losing their stability and high performance. On the other hand, the choice of electrolytes and fabrication methods is very important to really reach a commercial scale.

References

1. B.E. Conway, (Plenum Publishers, New York, 1999), p. 19
2. K.N. Hui, K.S. Hui, Z. Tang, V.V. Jadhav, Q.X. Xia, J. Power Sour. **330**, 195–203 (2016)
3. G. Oukali, E. Salager, M.R. Ammar, C.E. Dutoit, V.S. Kanian, P. Simon, E.R. Piñero, M. Deschamps, ACS Nano **13**, 12810–12815 (2019)
4. R. Sahoo, T.H. Lee, D.T. Pham, T.H.T. Luu, Y.H. Lee, ACS Nano **13**, 10776–10786 (2019)
5. Z. Chen, Y.C. Qin, D. Weng, Q.F. Xiao, Y.T. Peng, X.L. Wang, H.X. Li, F. Wei, Y.F. Lu, Adv. Funct. Mater. **19**, 3420–3426 (2009)
6. H.E. Becker, U. S. Patent 2, 800, 616, (to General Electric) (1957)
7. R. Kotz, M. Carlen, Electrochim. Acta **45**, 2483–2498 (2000)
8. R.A. Rightmire, U.S. Patent, 3288641, 29 Nov 1966
9. D.I. Boos, G. Heights, *Electrolytic Capacitor Having Carbon Paste Electrodes* (1970)
10. X. Li, J. Shao, S.K. Kim, C. Yao, J. Wang, Y.R. Miao, Q. Zheng, P. Sun, R. Zhang, P.V. Braun, Nature Comm. **9**, 2578 (2018)
11. P. Simon, Y. Gogotsi, Nature Mater. **7**, 845–854 (2008)
12. J.P. Zheng, J. Huang, T.R. Jow, J. Electrochem. Soc. **144**, 2026–2031 (1997)
13. H. Luo, G. Wang, J. Lu, L. Zhuang, L. Xiao, A.C.S. Appl, Mater. Interfaces **11**, 41215–41221 (2019)
14. A.S. Arico, P. Bruce, B. Scrosati, J.M. Tarascon, W.V. Schalkwijk, Nat. Mater. **4**, 366–377 (2005)

15. X. Chen, Y. Zhu, M. Zhang, J. Sui, W. Peng, Y. Li, G. Zhang, F. Zhang, X. Fan, ACS Nano **13**, 9449–9456 (2019)
16. N.R. Chodankar, S.J. Patil, G.S.R. Raju, D.W. Lee, D.P. Dubal, Y.S. Huh, Y.K. Han, Chem. Sus. Chem. **12**, 1–12 (2019)
17. W. Wei, X. Cui, W. Chen, D.G. Ivey, Chem. Soc. Rev. **40**, 1697–1721 (2011)
18. S.A. Sellam, Hashmi. ACS Appl. Mater. Interfaces. **5**, 3875–3883 (2013)
19. X. Sun, T. Xu, J. Bai, C. Li, A.C.S. Appl, Energy Mater. **2**, 8675–8684 (2019)
20. J. Wang, J. Wang, Z. Kong, K. Lv, C. Teng, Y. Zhu, Adv. Mater. **29**, 1703044 (2017)
21. N.M. Shinde, Q.X. Xia, P.V. Shinde, J.M. Yun, R.S. Mane, K.H. Kim, A.C.S. Appl, Mater. Interfaces **11**, 4551–4559 (2019)
22. B.E. Conway, V. Birss, J. Wojtowicz, J. Power Sour. **66**, 1–14 (1997)
23. M. Chuang, C.W. Huang, H. Teng, J.M. Ting, Energy Fuels **24**, 6468–6476 (2010)
24. J. Ge, B. Wang, J. Wang, Q. Zhang, B. Lu, Adv. Energy Mater. 2019, 1903277. https://doi.org/10.1002/aenm.201903277
25. H.J. Kang, Y.S. Huh, W.B. Im, Y.S. Jun, ACS Nano **13**(2019), 11935–11946 (1946)
26. P. Sharma, T.S. Bhatti, Energy Convers. Manag. **51**, 2901–2912 (2010)
27. I.S. Son, Y. Oh, S.H. Yi, W.B. Im, S.E. Chun, Carbon **159**, 283–291 (2020)
28. Q.Z. Zhang, D. Zhang, Z.C. Miao, X.L. Zhang, S.L. Chou, Small **14**, 1702883 (2018)
29. L. Yuan, X.H. Lu, X. Xiao, T. Zhai, J. Dai, F. Zhang, B. Hu, X. Wang, L. Gong, J. Chen, C. Hu, Y. Tong, J. Zhou, Z.L. Wang, ACS Nano **6**, 656–661 (2012)
30. L. Bao, J. Zang, X. Li, Nano Lett. **11**, 1215–1220 (2011)
31. G.G. Amatucci, F. Badway, A. Du Pasquier, T. Zheng, J. Electrochem. Soc. **148**, A930 (2001)
32. E. Frackowiak, Phys. Chem. Chem. Phys. **9**, 1774–1785 (2007)
33. J. Phiri, J. Dou, T. Vuorinen, P.A.C. Gane, T.C. Maloney, ACS Omega **4**, 18108–18117 (2019)
34. S. Liu, P. Yang, L. Wang, Y. Li, Z. Wu, R. Ma, J. Wu, X. Hu, Energy Fuels **33**, 6568–6576 (2019)
35. D.Y. Lee, D.V. Shinde, E.K. Kim, W. Lee, I.W. Oh, N.K. Shrestha, J.K. Lee, S.H. Han, Micropor. Mesopor. Mater. **171**, 53–57 (2013)
36. T. Kesavan, M. Partheeban, M. Vivekanantha, G. Kundu, M. Maduraiveeran, Sasidharan. Microporous Mesoporous Mater. **274**, 236–244 (2019)
37. H. Lin, Y. Liu, Z. Chang, S. Yan, S. Liu, S. Han, Microporous Mesoporous Mater. **292**, 109707 (2020)
38. C. Prehal, C. Koczwara, H. Amenitsch, V. Presser, O. Paris, Nature Comm. **9**, 4145 (2018)
39. Y. Koo, V.N. Shanov, S. Yarmolenko, M. Schulz, J. Sankar, Y, Yun. Langmuir **31**, 7616–7622 (2015)
40. I. Khakpour, A.R. Baboukani, A. Allagui, C. Wang, A.C.S. Appl, Energy Mater. **2**, 4813–4820 (2019)
41. D. Pang, M. Alhabeb, X. Mu, Y.D. Agnese, Y. Gogotsi, Y. Gao, Nano Lett. **19**, 7443–7448 (2019)
42. L. Yu, L. Hu, B. Anasori, Y.T. Liu, Q. Zhu, P. Zhang, Y. Gogotsi, B. Xu, ACS Energy Lett. **3**, 1597–1603 (2018)
43. D.A. Dikin, S. Stankovich, E.J. Zimney, R.D. Piner, G.H.B. Dommett, G. Evmenenko, S.T. Nguyen, R.S. Ruoff, Nature **448**, 457–460 (2007)
44. M.D. Stoller, S.J. Park, Y.W. Zhu, J.H. An, R.S. Ruoff, Nano Lett. **8**, 3498–3502 (2008)
45. W. Lv, D.M. Tang, Y.B. He, C.H. You, Z.Q. Shi, X.C. Chen, C.M. Chen, P.X. Hou, C. Liu, Q.H. Yang, ACS Nano **3**, 3730–3776 (2009)
46. D.W. Wang, F. Li, J.P. Zhao, W.C. Ren, Z.G. Chen, J. Tan, Z.S. Wu, I. Gentle, G.Q. Lu, H.M. Cheng, ACS Nano **3**, 1745–1752 (2009)
47. B.K. Kuila, B. Nandan, M. Bohme, A. Janke, M. Stamm, Chem. Commun. **47**, 5749–5751 (2009)
48. G.M. Suppes, B.A. Deore, M.S. Freund, Langmuir **24**, 1064–1069 (2008)
49. K. Zhou, Y. He, Q. Xu, Q. Zhang, A. Zhou, Z. Lu, L.K. Yang, Y. Jiang, D. Ge, X.Y. Liu, H. Bai, ACS Nano **12**, 5888–5894 (2018)
50. C. Pan, Z. Liu, W. Li, Y. Zhuang, Q. Wang, S. Chen, J. Phys. Chem. C **123**, 25549–25558 (2019)

51. Y. Gawli, A. Banerjee, D. Dhakras, M. Deo, D. Bulani, P. Wadgaonkar, M. Shelke, S. Ogale **6**, 21002 (2016)
52. M.O. Bamgbopa, J. Edberg, I. Engquist, M. Berggren, K. Tybrandt, J. Mater. Chem. A **7**, 23973–23980 (2019)
53. R. Wannapob, M. Yu. Vagin, I. Jeerapan, W. Cheung Mak, Langmuir **31**, 11904–11913 (2015)
54. C. Zhou, Y. Zhang, Y. Li, J. Liu, Nano Lett. **13**, 2078–2085 (2013)
55. C. Bora, J. Sharma, S. Dolui, J. Phys. Chem. C **118**, 29688–29694 (2014)
56. S. Yang, L. Sun, X. An, X. Qian, Carbohyd. Polym. **229**, 115455 (2020)
57. X. Zhang, Y. Lv, A. Liu, H. Li, D. Li, Z. Guo, J. Mu, Y. Wang, X. Liu, H. Che, J. Alloy. Compd. **820**, 153364 (2020)
58. S. Kulandaivalu, Y. Sulaiman, J. Power Sour. **419**, 181–191 (2019)
59. W. He, G. Zhao, P. Sun, P. Hou, L. Zhu, T. Wang, L. Lib, X. Xu, T. Zhai, Nano Energy **56**, 207–215 (2019)
60. H. Moon, H. Lee, J. Kwon, Y.D. Suh, D.K. Kim, I. Ha, J. Yeo, S. Hong, S.H. Ko, Sci. Rep. **7**, 1–10 (2017)
61. K. Jiang, I. A. Baburin, P. Han, C. Yang, X. Fu, Y. Yao, J. Li, E. Cánovas, G. Seifert, J. Chen, M. Bonn, X. Feng, X. Zhuang, Adv. Funct. Mater. (2019) 1908243
62. S. Luo, J. Zhao, J. Zou, Z. He, C. Xu, F. Liu, Y. Huang, L. Dong, L. Wang, H. Zhang, A.C.S. Appl, Mater. Interfaces **10**, 3538–3548 (2018)
63. C. Ma, W.T. Cao, W. Xin, J. Bian, M.G. Ma, Ind. Eng. Chem. Res. **58**(2019), 12018–12027 (2027)
64. Y. Song, H. Wang, W. Yu, J. Wang, G. Liu, D. Li, T. Wang, Y. Yang, X. Dong, Q. Ma, J. Power Sour. **405**, 51–60 (2018)
65. M.R. Kaiser, Z. Han, J. Wang, J. Power Sour. **437**, 226925 (2019)
66. X. Sun, Y. Zhang, J. Zhang, F. U. Zaman, L. Hou, C. Yuan, Energy Technol. (2019) 1901218
67. Y. Lv, M. Shang, X. Chen, P.S. Nezhad, J. Niu, ACS Nano **13**, 12032–12041 (2019)
68. J. Zhang, Y. Shi, Y. Ding, W. Zhang, G. Yu, Nano Lett. **16**, 7276–7281 (2016)
69. A. Lahiri, L. Yang, G. Li, F. Endres, ACS Appl. Mater. Interf. **11**, 45098–45107 (2019)
70. P. Geng, S. Cao, X. Guo, J. Ding, S. Zhang, M. Zheng, H. Pang, J. Mater. Chem. A **7**, 19465–19470 (2019)
71. X. Jia, C. Wang, C. Zhao, Y. Ge, G.G. Wallace, Adv. Funct. Mater. **26**, 1454–1462 (2016)
72. J. Liu, Xijun Xu, R. Hu, L. Yang, M. Zhu, Adv. Energy Mater. **6**, 1600256 (2016)
73. C. Zhu, P. Yang, D. Chao, X. Wang, X. Zhang, S. Chen, B.K. Tay, H. Huang, H. Zhang, W. Mai, H.J. Fan, Adv. Mater. **27**, 4566–4571 (2015)
74. I. Ryu, M.H. Yang, H. Kwon, H.K. Park, Y.R. Do, S.B. Lee, S. Yim, Langmuir **30**, 1704–1709 (2014)
75. V.K.A. Muniraj, P.K. Dwivedi, P.S. Tamhane, S. Szunerits, R. Boukherroub, M.V. Shelke, A.C.S. Appl, Mater. Interfaces **11**, 18349–18360 (2019)
76. D. Majumdar, T. Maiyalagan, Z. Jiang, Chem Electro Chem **6**, 4343–4372 (2019)
77. J. Zhang, J. Jiang, H. Lib, X.S. Zhao, Energy Environ. Sci. **4**, 4009–4015 (2011)
78. F. Cheng, J. Shen, W. Ji, Z. Tao, J. Chen, A.C.S. Appl, Mater. Interfaces **1**, 460–466 (2009)
79. F. Cheng, J. Zhao, W. Song, C. Li, H. Ma, J. Chen, P. Shen, Inorg. Chem. **45**, 2038–2044 (2006)
80. S. Devaraj, N. Munichandraiah, J. Phys. Chem. C **112**, 4406–4417 (2008)
81. O. Ghodbane, J.L. Pascal, F. Favier, A.C.S. Appl. Mater. Interfaces **1**, 1130–1139 (2009)
82. M. Toupin, T. Brousse, D.B. Elanger, Chem. Mater. **14**, 3184–3190 (2004)
83. V. Subramanian, H. Zhu, R. Vajtai, P.M. Ajayan, B. Wei, J. Phys. Chem. B **109**, 20207–20214 (2005)
84. J. Shao, W. Li, X. Zhou, J. Hu, Cryst Eng Comm **16**, 9987–9991 (2014)
85. M. Shen, S.J. Zhu, X. Liu, X. Fu, W.C. Huo, X.L. Liu, Y.X. Chen, Q.Y. Shan, H.C. Yaod, Y.X. Zhang, Cryst Eng Comm **21**, 5322–5331 (2019)
86. Z. Su, C. Yang, B. Xie, Z. Lin, Z. Zhang, J. Liu, B. Li, F. Kanga, C.P. Wong, Energy Environ. Sci. **7**, 2652–2659 (2014)
87. D.A. Tompsett, S.C. Parker, M.S. Islam, J. Am. Chem. Soc. **136**(4), 1418–1426 (2014)

88. A. Prasath, M. Athika, E. Duraisamy, A.S. Sharma, V.S. Devi, P. Elumalai, ACS Omega **4**, 4943–4954 (2019)
89. D. Moitra, C. Anand, B.K. Ghosh, M. Chandel, N.N. Ghosh, A.C.S. Appl, Energy Mater. **1**, 464–474 (2018)
90. J. Hao, S. Peng, H. Li, S. Dang, T. Qin, Y. Wen, J. Huang, F. Ma, D. Gao, F. Li, G. Cao, J. Mater. Chem. A **6**, 16094–16100 (2018)
91. V.D. Nithya, N.S. Arul, J. Mater. Chem. A **4**, 10767–10778 (2016)
92. M. Tian, R. Li, C. Liu, D. Long, G. Cao, A.C.S. Appl, Mater. Interfaces **11**, 15573–15580 (2019)
93. W. Huang, L. Li, D. Liang, W. Zhou, H. Wang, Z. Lu, S. Yu, Y. Fan, Inorg. Chem. Front. **6**, 2927–2934 (2019)
94. A. Pendashteh, S.E. Moosavifard, M.S. Rahmanifar, Y. Wang, M.F. El-Kady, R.B. Kaner, M.F. Mousav, Chem. Mater. **27**, 3919–3926 (2015)
95. L. Kumar, H. Chauhan, N. Yadav, N. Yadav, S.A. Hashmi, S. Deka, A.C.S. Appl, Energy Mater. **1**, 6999–7006 (2018)
96. S. Zhou, W. Wei, Y. Zhang, S. Cui, W. Chen, L. Mi, Scientific Reports **9**, 12727 (2019)
97. S. Anandan, B.G.S. Raj, G.J. Lee, J.J. Wu, Mater. Res. Bull. **48**, 3357–3361 (2013)
98. J. Kang, J. Sheng, Y. Ji, H. Wen, X.Z. Fu, G. Du, R. Sun, C.P. Wong, Chemistry Select **2**, 9570–9576 (2017)
99. M. Gao, Z.Y. Guo, X.Y. Wang, W.W. Li, Chemsuschem **12**, 5291–5299 (2019)
100. J. Azadmanjiri, Vijay K. Srivastava, P. Kumar, M. Nikzad, J. Wang, A. Yu, J. Mater. Chem. A, **6**, 702–734 (2018)
101. M. Hu, C. Cui, C. Shi, Z.S. Wu, J. Yang, R. Cheng, T. Guang, H. Wang, H. Lu, X. Wang, ACS Nano **13**, 6899–6905 (2019)
102. N. Maheswari, G. Muralidharan, Energy Fuels **29**, 8246–8253 (2015)
103. X. Zhang, X. Wang, L. Jiang, H. Wu, C. Wu, J. Su, J. Power Sources **216**, 290–296 (2012)
104. C.Y. Foo, A. Sumboja, D.J.H. Tan, J. Wang, P.S. Lee, Adv. Energy Mater. **4**, 1400236 (2014)
105. B. Pal, S. Yang, S. Ramesh, V. Thangadurai, R. Jose, Nanoscale Adv. **1**, 3807–3835 (2019)
106. M. Armand, F. Endres, D.R. Macfarlane, H. Ohno, B. Scrosati, Nat. Materials **8**, 621–629 (2009)
107. F. Beguin, V. Presser, A. Baldacci, E. Frackowiak, Adv. Mater. **26**, 2219–2251 (2014)
108. A. Kundu, T.S. Fisher, A.C.S. Appl, Energy Mater. **1**, 5800–5809 (2018)
109. X. Yang, F. Zhang, L. Zhang, T. Zhang, Y. Huang, Y. Chen, Adv. Funct. Mater. **23**, 3353–3360 (2013)
110. D. Lee, Y.H. Song, U.H. Choi, J. Kim, A.C.S. Appl, Mater. Interfaces **11**, 42221–42232 (2019)
111. M.D. Stoller, R.S. Ruoff, Energy Environ. Sci. **3**, 1294–1301 (2010)
112. X. Lang, A. Hirata, T. Fujita, M. Chen, Nat. Nanotechnol. **6**, 23–2362 (2011)
113. B. Qin, Y. Han, Y. Ren, D. Sui, Y. Zhou, M. Zhang, Z. Sun, Y. Ma, Y. Chen, Energy Technol. **6**, 306–311 (2018)
114. H.V. Helmholtz, Ann. Phys. (Leipzig) **89**, 211–233 (1853)
115. G. Gouy, J. Phys. **4**, 457–468 (1910)
116. D.L. Chapman, Philos. Mag. **6**, 475–481 (1913)
117. O.Z. Stern, Electrochem. **30**, 508–516 (1924)
118. E. Gileadi, B.E. Conway, J. Chem. Phys. **39**, 3420–3430 (1963)
119. E. Frackowiak, F. Beguin, Carbon **39**, 937–950 (2001)
120. C.C. Hu, K.H. Chang, M.C. Lin, Y.T. Wu, Nano Lett. **6**, 2690–2695 (2006)
121. H. Zhang, G. Cao, Z. Wang, Y. Yang, Z. Shi, Z. Gu, Nano Lett. **8**, 2664–2668 (2008)
122. D. Choi, G.E. Blomgren, P.N. Kumta, Adv. Mater. **18**, 1178–1182 (2006)
123. I.W.P. Chen, Y.C. Chou, P.Y. Wang, J. Phys. Chem. C **123**, 17864–17872 (2019)
124. Y. Liu, L. Cao, J. Luo, Y. Peng, Q. Ji, J. Dai, J. Zhu, X. Liu, ACS Sustain. Chem. Eng. **7**, 2763–2773 (2019)
125. S. Sharma, R. Soni, S. Kurungot, S.K. Asha, J. Phys. Chem. C **123**, 2084–2093 (2019)
126. Y. Chen, W.K. Pang, H. Bai, T. Zhou, Y. Liu, S. Li, Z. Guo, Nano Lett. **17**, 429–436 (2017)
127. D.W. Kim, S.M. Jung, H.Y. Jung, J. Mater. Chem. A **8**, 532–542 (2020)

Chapter 3
Basics of Ferrites: Structures and Properties

3.1 Introduction

The word 'ferrite', derived from the Latin 'ferrum' means iron. The first type of magnetic material known to human was lodestone which is made up of the ore magnetite (Fe_2O_3). Ferrite is a ceramic material made by mixing iron oxide with a metallic element. They are electrically non-conducting insulators. This is believed that there ferrites were initially discovered in ancient Greece around the time period of 800 BC. Ferrite is a class of magnetic oxide compound that contains iron oxide as a principal component. In 1928, Forestier prepared ferrites by a heat treatment technique [1]. In 1947, several ferrites were developed as commercially important materials by Snoek [2]. The major booming of the ferrite industry was led by the large-scale introduction of television in the 1950s. In television processing, ferrite cores are used as high-voltage transformers and electron beam deflection yokes due to their strong magnetic properties, high electrical resistivity, low magnetic losses, etc. [3].

3.2 Classification of Ferrites

Ferrites are classified according to their crystal structures and magnetic properties as follows (Fig. 3.1).

© The Authors 2020
V. V. Jadhav et al., *Bismuth-Ferrite-Based Electrochemical Supercapacitors*,
SpringerBriefs in Materials, https://doi.org/10.1007/978-3-030-16718-9_3

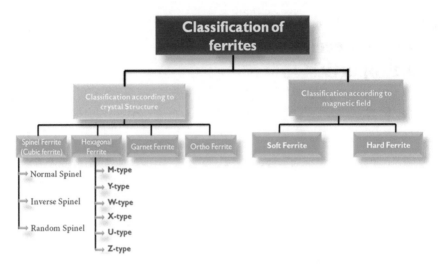

Fig. 3.1 Classification of ferrites

3.2.1 Based on Crystal Structure

3.2.1.1 Spinel Ferrites (Cubic Ferrites)

Spinel ferrites are characterized by the formula MFe_2O_4, where M stands for divalent metal ions like Cu, Ni, Mg, Mn, Co, Zn, Cd, etc. The Fe^{3+} can be replaced by other trivalent ions like Al, Cr, Ga, In, etc. Spinel ferrites is the most important class of magnetic materials that exhibit a number of interesting applications. Spinel ferrites are magnetically soft and are alternative to metallic magnets such as Fe and layered Fe–Si alloys, but exhibit enhanced performance due to their outstanding magnetic properties [1, 4]. They also can provide high electrical resistivity and low magnetic losses. The two popular ceramic magnets—nickel–zinc ferrites and manganese–zinc ferrites—are the major members of the spinel ferrite family. They are fascinating ceramic materials due to their high electrical resistivity, high magnetic permeability, and possible modification of intrinsic properties over a wide spectrum [5]. Spinel ferrite crystallizes in the cubic structure and is composed of a close-packed oxygen anions arrangement in which 32 oxygen ions form the unit cell. These anions are packed in a face-centered cubic (FCC) arrangement leaving two kinds of spaces between anions: tetrahedrally coordinated sites (A), surrounded by four nearest oxygen atoms, and octahedrally coordinated sites (B), surrounded by six nearest-neighbor oxygen atoms. The crystal structure is cubic with spinel-type, i.e., $MgAl_2O_4$ with two interstitial sites, namely tetrahedral (A) and octahedral (B). The unit cell is made up of eight units (cubes) and thus can be written as $M_8Fe_{16}O_{32}$ which was initially determined by Bragg and Nishikawa [6]. The spinel ferrite structure is shown in Fig. 3.2.

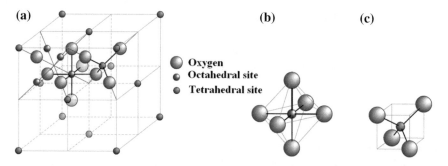

Fig. 3.2 **a** Spinel ferrite unit cell structure, **b** octahedral interstice, and **c** tetrahedral interstice

In Fig. 3.2a, the ionic positions are the same in octants sharing only one edge and different in octants sharing a face. Each octant contains four oxygen ions. In Fig. 3.2a, ionic positions at only two adjacent octants are shown, where the octant on the left contains octahedral and the one on the right contains tetrahedral sites. All ions are positioned along the body diagonals of the octants, and the octant on the right contains a tetrahedral site at the octant center.

(a) *Classification*

On the basis of the distribution of cations in the two principal sites, tetrahedral site (A) and octahedral site (B) [7], spinel ferrites are classified into three types as follows:

(i)   Normal spinel ferrites

A ferrite is called normal spinel when the divalent metal ions occupy the tetrahedral (A) site while $2Fe^{3+}$ ions are at octahedral (B) site. Normal spinel structures are commonly cubic close-packed oxides with eight tetrahedral and four octahedral sites per formula unit. The best examples of normal spinel ferrites are zinc ($ZnFe_2O_4$) and cadmium ferrites ($CdFe_2O_4$), where the divalent metallic ions Zn^{2+} or Cd^{2+} are at the (A) site, while Fe^{3+} ions are at (B) site. The cation distribution can be, in general, represented as $(M)^A [Fe_2]^B O_4$.

(ii)   Inverse spinel ferrites

In inverse spinel ferrite, one trivalent ferric ion Fe^{3+} is at the tetrahedral (A) site while the remaining trivalent ferric ions Fe^{3+} and the divalent metallic ions M^{2+} are at the (B) site. In fact, most of the simple ferrites, e.g., ferric oxide (Fe_3O_4), nickel ferrite ($NiFe_2O_4$), and cobalt ferrite ($CoFe_2O_4$) are of this category [8]. The cation distribution in inverse spinel ferrite is presented as $(Fe)^A [M\ Fe]^B O_4$.

(iii)   Random spinel ferrites

The divalent metal ions M^{2+} and trivalent Fe^{3+} ions are distributed at both tetrahedral (A) site and octahedral (B) site then the ferrite is termed as random spinel ferrite. The best known example of the random spinel ferrite is copper ferrite ($CuFe_2O_4$). The distribution of ions between two types of sites is determined by a delicate balance of contributions, such as the magnitude of ionic radii, their electronic configuration, and the electrostatic energy of the lattice [9, 10]. In general, the cation distribution in random spinel is given by $(M_{1-x}Fe_x)^A [M_x Fe_{2-x}]^B O_4$.

3.2.1.2 Hexagonal Ferrites

Hexaferrite was first identified by Went, Rathenau, Gorter, and Van Oostershout 1952 [11], and Jonker, Wijn, and Braun 1956. Hexagonal ferrites are also called as rhombohedral ferromagnetic oxides. Hexagonal ferrites are generally denoted by an $MFe_{12}O_{19}$ formula, where M is an element like barium (Ba), strontium (Sr), calcium (Ca), or lead (Pb). In these types of ferrites, oxygen ions avail a closed packed hexagonal crystal structure and have a high coercively with permanent (hard) magnet feature [12]. Hexagonal ferrites are referred to as hard since the direction of magnetization cannot be changed easily to another axis. The examples of hexaferrite are barium ferrite ($BaFe_{12}O_{19}$) and strontium ferrite ($SrFe_{12}O_{19}$). Hexagonal ferrite has become extremely important materials commercially and technologically. It is used as permanent magnets, most common applications are as magnetic recording, data storage materials, and as components in electrical devices, mostly those operating at microwave/GHz frequencies [13]. Jang et al. studied Enhanced Switchable Ferroelectric Photovoltaic Effects in Hexagonal Ferrite Thin Films via Strain Engineering [14]. Figure 3.3 reveals the structure of the $BaFe_{12}O_{19}$ hexaferrite. In Fig. 3.3, the arrows over Fe ions represent the direction of spin polarization. $2a$, $12\,k$, and $4f_2$ are octahedral, $4f_1$ are tetrahedral, and $2b$ are hexahedral (trigonal bipyramidal) sites. The unit cell contains a total of 38 O^{2-} ions, 2 Ba^{2+} ions, and 24 Fe^{3+} ions. The Fe^{3+} ions in $12\,k$, $2a$, and $2b$ sites (16 total per unit cell) have their spins up, while the Fe^{3+} ions in $4f_1$ and $4f_2$ sites (8 total per unit cell) have their spins down, resulting in a net total of 8 spins up, and therefore, a total moment of $8 \times 5\,\mu B = 40\,\mu B$ per unit cell that contains two Ba^{2+} ions.

The R and S subunits endow $R = (Ba^{2+}Fe_6^{3+}O_{11}^{2-})^{2-}$ and $S = (Fe_6^{3+}O_8^{2-})^{2+}$ formulae. The asterisk (*) indicates that the corresponding subunit is rotated 180° around the hexagonal axis.

Types:

Hexaferrites are classified into six subclasses as shown in Table 3.1, where, a can be barium (Ba), strontium (Sr), calcium (Ca), or lead (Pb), and Me is a divalent transition metal ion. Table 3.1 confirms the types of hexaferrites.

Recently, hexaferrite has been explored for more exotic application such as electronic components for mobile, wireless communication, electromagnetic wave absorbers for EMC, RAM. Also, hexaferrite materials are common used in common mode chokes, electromagnetic interference filters, current sensors, handheld devices, spike suppression and gate drive transformers, shield beads, snap-on cores, flat cable beads, automotive industry, and consumer goods. Also they are used in computers and peripherals, communication systems, automobiles, switch mode power supplies DC-DC converters, ignition coils, etc. [15].

Fig. 3.3 Schematic structure of $BaFe_{12}O_{19}$ hexaferrite

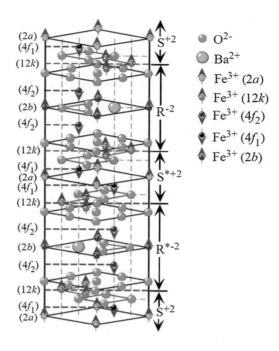

Table 3.1 Types of hexaferrites

Sr. no.	Hexaferrite	Chemical formula
1	M-type	$AFe_{12}O_{19}$
2	Y-type	$A_2Me_2Fe_{12}O_{22}$
3	W-type	$AMe_2Fe_{16}O_{27}$
4	X-type	$A_2Me_2Fe_{28}O_{46}$
5	U-type	$A_4Me_2Fe_{36}O_{60}$
6	Z-type	$A_3Me_2Fe_{24}O_{41}$

3.2.1.3 Garnet Ferrites

Recently, garnet-based nanoferrites are being focused by the researchers because of their extensive applications in various fields. Garnets are basically minerals and represented by a general formula $X_3Fe_5O_{12}$, containing two magnetic ions, one of which is typically being iron and another is rare earth, where 'X' is rare earth element like Sm, Eu, Gd, Tb, Dy, Er, Tm, Lu, and Y. Garnets have dodecahedral (12-coordinated) sites in addition to tetrahedral and octahedral sites which are magnetically hard. Bertaut and Forret prepared [11] $X_3Fe_5O_{12}$ in 1956 and measured their magnetic properties. In 1957, Geller and Gilleo prepared $Gd_3Fe_5O_{12}$ and investigated ferromagnetic properties [3]. Their unit cell shape is cubic, and the edge length is about 12.5 Å. They found complex crystal structure which is important due to their applications in memory devices. Yttrium iron garnet $Y_3Fe_5O_{12}$ is a well-known garnet.

The garnets have orthorhombic crystal structure (oxygen polyhedral, surrounding the cations) but with trivalent cations (including rare earth and Fe^{3+}) occupying tetrahedral (d), octahedral (a), or dodecahedral-a 12-sided distorted polyhedral (c) sites. Specifically, the interaction between tetrahedral and octahedral sites is antiparallel, and the net magnetic moment is antiparallel to the rare earth ions on the 'c' sites. The garnet structure is one of the most complicated crystal structures, and it is difficult to draw a two-dimensional representation, showing all the ions, i.e., 160 in the unit cell. For simplicity, only an octant of a garnet structure that shows just the cation positions is shown in Fig. 3.4. The garnet structure is composed of a combination of octahedral (trivalent cation surrounded by six oxygen ions), tetrahedral (trivalent cations surrounded by four oxygen ions), and 12-sided polyhedral dodecahedral (trivalent cations surrounded by 8 oxygen atoms) sites, the orientations of which are shown in Fig. 3.5 [16].

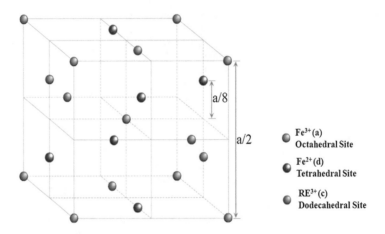

Fig. 3.4 Schematic octant of a garnet crystal structure with only cation positions

Fig. 3.5 An octant of a garnet crystal structure

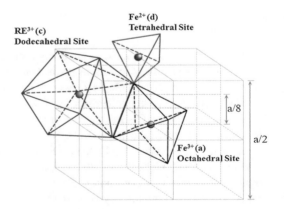

Fig. 3.6 Perovskite
structure of an orthoferrite

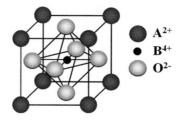

A^{2+}
B^{4+}
O^{2-}

3.2.1.4 Orthoferrites

Orthoferrites are represented with $MFeO_3$, where M is one or more rare earth elements. Orthoferrites have an orthorhombic crystal structure and are weakly ferromagnetic in nature. The praseodymium orthoferrite ($PrFeO_3$), lanthanum orthoferrite ($LaFeO_3$), and dysprosium orthoferrite ($DyFeO_3$) are examples of orthoferrite materials. Orthoferrites are used in communication techniques, optical Internet, cathodes in solid oxide fuel cells, catalysts, gas separators, spin valves, magneto-optic materials, sensors of magnetic fields, and electrical circuits [17]. The perovskite structure is shown in Fig. 3.6. Large divalent or trivalent ions (A) occupy the corners of a cube, and small trivalent or tetravalent metal ions (B) occupy the center of the cube. The oxygen ions are situated centrally on the faces of the cube.

3.2.2 According to Magnetic Properties

3.2.2.1 Soft Ferrites

Soft ferrite is an iron-oxide-based soft magnetic material. The word 'soft' means temporary in the sense that the ferromagnetism emerges only when a magnetic field is applied. All magnetic elements in the pure form are soft. Soft ferrites are ferrimagnetic in nature with cubic crystal structure and they are characterized by chemical formula $MO \cdot Fe_2O_3$, where M is a transition metal ion like iron, nickel, manganese, and zinc. Manganese–zinc ferrites act as soft magnets up to high frequencies of 10 MHz which are easily magnetized and demagnetized, so store or transfer magnetic energy in alternating or other changing waveforms. Soft ferrites are used for transformer cores, switch mode power supplies, inductors, convertors, electromagnetic interference filters, picture tubes, in electronics, biological application, and yokes. They are being extensively used in television, telecommunication, space research and military devices and include variety of industrial applications [18–22].

Table 3.2 Comparative properties of soft and hard magnetic ferrites

Soft ferrite	Hard ferrite
High saturation magnetization (1-2T)	High saturation magnetization (0.3-6T)
Low coercivity	High coercivity
High permeability	Not important, but low
Low anisotropy	High anisotropy
Low magnetostriction	Not important
High Curie temperature	High Curie temperature
Low losses	High-energy product
High electrical resistivity	Not important

3.2.2.2 Hard Ferrites

Hard ferrites are well-known magnetic materials that are generally used for the production of ceramic magnets. Hard ferrite magnets are produced with iron oxide and barium or strontium oxide. It displays ferromagnetism in the absence of an external field. Permanent ferrite magnets are made up of hard ferrites, which have a high coercively and a high remanence after magnetization. These ferrites are composed of iron and barium or strontium oxides. In a magnetically saturated state, they conduct magnetic flux well and have a high magnetic permeability. This enables them to store stronger magnetic fields than iron itself. They are cheap and are widely used in household products such as refrigerator magnets. The hexagonal ferrite structure is found in both $BaO \cdot 6Fe_2O_3$ and $SRO \cdot 6Fe_2O_3$, but Sr-M hexaferrite has slightly superior magnetic properties. Hard ferrites are found in microwave devices, recording media, magneto-optic media, and telecommunication and electronic industries [13, 23]. They are also used in transformer core, cathode ray tubes, small DC motors, compact torque devices, and magnetic latches [24]. The difference between soft ferrite and hard ferrites is given in Table 3.2.

3.3 Conclusions

The fast growth of research on magnetic materials has been revolutionizing in the field of energy storage devices. This chapter explains basic ferrite properties and different types of ferrites. It also discusses the types of ferrites under classification them. Ferrite materials are extensively applied for energy storage applications such as supercapacitor and batteries. Recently, mixed bismuth ferrites, spinel bismuth, cobalt, and nickel ferrites are of great interest for supercapacitors. These materials will be active electrode materials in future useful in energy devices because of their natural abudancy, low-cost, and ecofriendly nature.

References

1. E.C. Snelling, *Soft Ferrites: Properties and Applications*, 1st edn. (Iliffe Books Ltd., London, 1969)
2. J.L. Snoek, *New Developments in Ferromagnetic Materials* (Elsevier, Amsterdam, The Netherlands, 1947)
3. J. Smit, H.P.J. Wijn, *Ferrites* (Philips' Technical Library, Eindhoven, The Netherlands, 1959)
4. M. Sugimoto, J. Amer. Cer. Soc. **82**, 269–280 (1999)
5. L.L. Hench, J.K. West, Chem. Rev. **90**, 33–72 (1990)
6. W.H. Bragg, Phil. Mag. **30**, 305–315 (1915)
7. P.I. Slick, *Ferrites for Non-Microwave Applications*, vol. 2. (1980), pp. 189–241
8. E.J.W. Verwey, E.L. Heilman, Chem. Phys. **15**, 174–180 (1947)
9. J.L.O. Quinnonez, U. Pal, M.S. Villanueva, ACS Omega **3**, 14986–15001 (2018)
10. G. Hu, J.H. Choi, C.B. Eom, V.G. Harris, Y. Suzuk, Phy. Rev. B **62**, 779–782 (2000)
11. K.J. Standley, *Oxide Magnetic Materials*, 2nd edn. (Oxford University Press, 1972), p. 254
12. A.L. Stuijts, G.W. Rathenau, G.H. Weber, Philips Tech. Rev. **16**, 141 (1954)
13. R.C. Pullar, Prog. Mater. Sci. **57**, 1191–1334 (2012)
14. H. Han, D. Kim, K. Chu, J. Park, S.Y. Nam, S. Heo, C.H. Yang, H.M. Jang, A.C.S. Appl, Mater. Interfaces **10**, 1846–1853 (2018)
15. D. Makovec, B. Belec, T. Gorsak, D. Lisjak, M. Komelj, G. Drazicd, S. Gyergyek, Nanoscale **10**, 14480–14491 (2018)
16. S. Geller, M.A. Gilleo, J. Phys. Chem. Solids **3**, 30–36 (1957)
17. H. Xu, X. Hu, L. Zhang, Cryst. Growth Des. **8**, 2061–2065 (2008)
18. L. Zivanov, M. Damnjanovi, N. Blaz, A. Maric, M. Kisic, G. Radosavljevic, Mag., Ferro. Multif. Metal Oxides (2018) 387–409
19. C. Pahwa, S.B. Narang, P. Sharma, J. Alloy. Compd. **815**, 152391 (2020)
20. S. Debnath, R. Das, J. Mol. Struct. **1199**, 127044 (2020)
21. S. He, H. Zhang, Y. Liu, F. Sun, X. Yu, X. Li, Li Zhang, L. Wang, K. Mao, G. Wang, Y. Lin, Z. Han, R. Sabirianov, H. Zeng, Small **14**, 1800135 (2018)
22. A. Vermaa, M.I. Alama, R. Chatterjeea, T.C. Goelb, R.G. Mendiratta, J. Mag. Mag. Mater. **300**, 500–505 (2006)
23. R.C. Pullar, ACS Combinatorial Sci. **14**, 425–433 (2012)
24. S.E. Kushnir, D.S. Koshkodaev, P.E. Kazin, D.M. Zuev, D.D. Zaytsev, M. Jansen, Adv. Eng. Mater. **16**, 884–888 (2014)

Chapter 4
Bismuth Ferrites: Synthesis Methods and Experimental Techniques

This chapter is mainly concerned with the synthesis methods available for preparing bismuth ferrites either in powders or in thin-films. The experimental techniques employed for their characterizations are thoroughly explored. The synthesis methods used for obtaining ferrites of different structures and morphologies including hydrothermal, solvothermal, microwave, mechanothermal, seed-hydrothermal, sol-gel, chemical bath deposition, so-called coprecipitation or wet chemical synthesis, radio frequency, thermal plasma, spray pyrolysis, and electrodeposition methods are briefed. Theoretical details of analytical techniques used for physical measurements including field-emission scanning electron microscopy (FE-SEM), energy-dispersive X-ray analysis (EDX), surface wettability, X-ray diffraction (XRD), and Raman shift analysis are also highlighted. The theoretical measurements employed in electrochemical supercapacitors (cyclic-voltammetry, galvanostatic charge/discharge, electrochemical impedance spectroscopy, etc.) are discussed at the end.

4.1 Synthesis Methods

Synthesis methods play an important role in controlling the size and surface area of ferrite products. A number of chemical methods, including thermal methods, sol-gel, solid-state reactions, and coprecipitation, are effectively being used, in literature, to prepare ferrite products in different structures and morphologies.

© The Authors 2020
V. V. Jadhav et al., *Bismuth-Ferrite-Based Electrochemical Supercapacitors*,
SpringerBriefs in Materials, https://doi.org/10.1007/978-3-030-16718-9_4

4.1.1 Thermal Synthesis

Many developments in the thermal chemical method to synthesize bismuth ferrite nanostructures have been made using a variety of chemical approaches, including hydrothermal [1–11], solvothermal [12–15], microwave [16–18], mechanothermal [19, 20], and seed-hydrothermal [21]. In each of these methods, an iron salt usually $Fe(NO_3)_3.9H_2O$ or $FeCl_3.6H_2O$ is used. Salt is dissolved in water or another solvent under stirring, and depending on the metal salt, pH ~7–12 is adjusted. The mixture is placed into an autoclave and heated for 12–24 h, followed by naturally cooling to room temperature. The solid is collected by filtration or centrifugation and washed, followed by drying at approximately 85 °C overnight. For mechanothermal treatment, rather than dissolving, the precursor compounds are grounded in a ball-mill using the same basic procedure. Seed-hydrothermal method uses a seed of the metal oxide (M_2O_3) with the Fe (III) salt. The two compounds are placed into an autoclave and heated using the same procedure as described above.

4.1.2 Sol-Gel

The sol-gel method developed in the 1960's mainly is mainly useful in the nuclear industry. In recent years, it is one of the well-established synthesis approaches to prepare different inorganic materials such as bismuth ferrite, cobalt ferrite, and nickel ferrite. Though the sol-gel method has made an impact on material technology to its scale up potential in preparing conventional materials, this method has potential control over the surface properties and textural of the materials. Sol-gel and citrate methods involve the addition of the metal and iron precursors, along with citric acid, to form a gel. The precursors are dissolved in water or ethanol and stirred vigorously at pH ~9 until a gel-like material is formed. Citric acid or tartaric acid assists in the homogeneous distribution of the metal ions into solution. The gel is dried and then sintered at temperatures ranging from 450 to 800 °C at various periods of time. In this method, the formation of a gel provides a high degree of homogeneity and reduces the need for atomic diffusion during the solid-state calcinations [22]. The sol-gel technique is a cheap and low-temperature method that allows for the fine control over the products and their chemical compositions.

The sol-gel method is classified into three categories which are as follows:

(a) Spray coating

Spray coating is one of the most important techniques of the surface modification method. In this technique, the thin films are deposited by spraying various volumes of spray solution at room temperature and high temperature. This technique can provide thick coating with high deposition rate as comparative to other deposition technique.

(b) Dip coating

Dip coating is one of the most effective processes which is used in industry to manufacture bulk products such as coated fabrics, condoms, and in the biomedical field. It is also used academic nanomaterial science and engineering to synthesize thin-film electrode.

Dip coating consists of following steps;

(i) immersion of substrate into a tank containing coating material,
(ii) removal of the substrate from the tank, and
(iii) drying of films.

Dip coating is a popular way of producing a thin and uniform coating onto flat or cylindrical substrates.

(c) Spin coating

Spin coating technology is a worldwide and intensively used process for applying uniform thin films on flat substrates. An excess amount of solution is placed on the substrate, which is then rotated at high speed in order to spread the fluid by centrifugal force. A machine used for spin coating is called a spin coater. Rotation is continued while the fluid spins off the edges of the substrate, until the desire thickness of the films is achieved. The thickness of the films also depends on the concentration of the solution and the solvent used. It is one of the easiest ways to deposit various types of materials that are in solution form.

Nowadays, many phases of bismuth ferrite thin films are being synthesized by a sol-gel method [23–29]. Sol-gel-derived materials have diverse applications in optics, electronics, energy, space, etc. [30].

4.1.3 Chemical Bath Deposition

Recently, chemical bath deposition (CBD) is of greater commercial value than either thermal evaporation or sputtering and has attracted the attention of researchers due to its simplicity, least cost, convenience, uniform, stable, adherent, repeatability, large area scaling, and commercial production capability [31]. This method is used to deposit thin film and nanomaterial and is first described in 1863. It is simplest method that requires only solution containers and substrate mounting device. In this method, thin-film growth depends on time, temperature, pH, composition of the solution, and nature of the substrate.

In 1933, Bruckman deposited lead (II) sulfide (PbS) thin film by CBD method which in 1946 was used in infrared applications. The basic principle involved in the synthesis of thin films/powders by the CBD is the controlled precipitation of the desired compound from a solution of its constituents. The ionic product must exceed the solubility product; thus, the formation of thin films on substrate by ion-by-ion condensation [32]. The CBD is one of the cheapest methods used to deposit thin ferrite films [33–35] and nanomaterials, as it does not require expensive sophisticated equipments which is a scalable and can be employed for large area batch processing or continuous deposition. The CBD involves two steps—nucleation and particle growth—and is based on the formation of a solid phase from a solution. In the CBD procedure, the substrate is immersed in an aqueous solution containing the mixture of precursors. The major advantage of CBD is that it requires only solution containers/beakers and substrate on which sample is to be deposited. The one drawback of this method is the wastage product/solution after every deposition. Among various deposition techniques, CBD yields stable, adherent, uniform, and hard films with good reproducibility by a relatively simple operation steps. It is one of the scalable methods for preparing highly efficient thin films in a simple manner. The growth of thin films strongly depends on growth conditions, such as duration of deposition, composition and temperature of the solution, and topographical and chemical nature of the substrate. Using CBD technique, $NiZnFe_2O_4$ [34], $NiFe_2O_4$ [35], $CoFe_2O_4$ [36], and $BiFeO_3$ [37] ferrite thin films have been synthesized in literature.

4.1.4 Co-precipitation

Coprecipitation is a very convenient and facile way to synthesize ferrite nanoparticles from aqueous salt solutions at room temperature or at elevated temperature. This method is used to prepare ferrites nanomaterials similar to thermal methods [38–41]. Coprecipitation reactions involve the formation of simultaneous nucleation, growth, coarsening, and agglomeration processes to take place. In this method, Fe (III) and metal salts are dissolved in water along with a surfactant, oleic acid, under stirring, and gentle heating. The reaction mixture pH is increased to 7–10 in order to precipitate the ferrite particles. The solid particles are then filtered and washed with distilled water or ethanol and subsequent dried at 80–100 °C. Once drying is completed, the particles are then air-heated at various temperatures to determine the effects of the temperature on the activity of the particles.

4.1.5 Solid-State Reaction

Solid-state synthesis methods are the most widely used. This method involves mixing of raw materials by either grinding or melting the starting materials together or simply applying heat to a mixture of starting materials. It can take place with both wet and dry processes in the absence of solvents. An aqueous suspension, agitator mill, or vibration drum is used in wet mixing methods. This method is extremely effective but requires energy for dewatering and drying. Dry mixing is done either by grinding and mixing in a drum or ball mill. In the solid-state route [42], Bi_2O_3 and Fe_2O_3 are reacted at a temperature in the range of 800–830 °C in nitrogen and oxygen environments, and unreacted $Bi_2O_3/Bi_2Fe_4O_9$ phases are removed by washing in nitric acid. Solid-state reactions involve heat-treating powders of the iron and metal salts. For example, to synthesize $CaFe_2O_4$, Fe_2O_3 and $CaCO_3$ powders are mixed and then heated up to 1000 °C for 24 h [43, 44]. The disadvantage of this process lies in the necessity of leaching the unwanted phases using an acid and effectively providing coarser powder and also the reproducibility of the process is quite poor.

4.1.6 Radio-Frequency Thermal Plasma

This thermal plasma is an area of the arc discharge. It operates at the low-pressure environment or atmospheric pressure. It generates 10,000° plasma from various gases using electromagnetic induction. This method introduces materials such as powders, gases, and fluids into the plasma. It is possible to induce vaporization, melting, dissociation, and chemical reactions for applications such as nanoparticle synthesis, film formation, powder spheroidization and reaction, and dissociation of harmful gases. A radio-frequency plasma method consists of a TEKNA PL50-type plasma torch head with a 50 kW and 3 MHz power supply stations and a gas expansion reactor chamber connected to a filtering unit for producing the ferrite powder [45, 46]. Argon is used as plasma gas, and the sheath gas consists of a mixture of Ar and H_2. Precursor powder is injected through the plasma stream using Ar, as a carrier gas. After the plasma ignited, air can be introduced into the reaction chamber as a source of oxygen. Then, the final powders from both the reaction chamber and the dry filter are collected and analyzed. More detailed processing procedures are described in references [47–49].

4.1.7 Pulsed Laser Deposition

Pulsed laser deposition (PLD) is a versatile deposition method used to grow thin films of a large number of materials with technological interest, such as carbides, borides, and nitrides. It is also used for the production of superconducting and insulating component to improved wear and biocompatibility for medical application. The PLD method is a physics vapor deposition technique where high-power pulse laser beam is focused inside a vacuum chamber to strike a target of the material that is to be deposited. It is a well-established method for growing ferrite films [50–54]. In conventional PLD, laser pulses from a high energy laser ablate a homogeneous target forming a molecular flux. The substrate intercepts the plume allowing for film growth on selected, often lattice-matched substrates. The PLD has been used in the deposition of garnet, spinel, and hexaferrite ferrites. The deposition of epitaxial yttrium iron garnet films by PLD was first demonstrated by Dorsey et al. in 1993 [55, 56]. Films with a thickness of about 1 mm were deposited on (111) gadolinium gallium garnet substrates. Thin films of $Bi_3Fe_5O_{12}$ (BIG), $Eu_1Bi_2Fe_5O_{19}$ (EBIG) as well as YIG/BIG ($Y_3Fe_5O_{12}/Bi_3Fe_5O_{12}$) and YIG/EBIG heterostructures were synthesized by Simion et al. [57].

4.1.8 Spray Pyrolysis

This is a simple and inexpensive method for producing various materials, including both thin-and thick-film forms. Unlike many other film deposition methods, spray pyrolysis is a very simple and relatively cost-effective processing method. Typical spray pyrolysis equipment consists of an atomizer, precursor solution, substrate heater, and temperature controller. In this process, thin films are deposited by spraying a solution on a preheated substrate surface, where the constituents react to form a chemical compound. The preparation of mesoporous $NiFe_2O_4$ spheres [58], $ZnFe_2O_4/Fe_2O_3$ composite [59], $LaFeO_3$ [60], cobalt ferrite [61], and magnesium ferrite [62] thin film by spray pyrolysis has been noted in literature. Deposition and advances in nanostructures fabricated visa spray pyrolysis and their application in energy storage and conversion is reported in the review article [63]. Li et al., synthesized chainlike MFe_2O_4 (M = Cu, Ni, Co, and Zn) nanoaggregates by facile flame spray pyrolysis technique for reduction of nitroaromatic compounds [64].

4.1.9 Electrodeposition

Electrodeposition is a well-known conventional surface modification technique to improve the surface characteristics, decorative and functional, of a wide variety of materials. It is a versatile and inexpensive technique can be performed at mild temperatures ($\leq 100\ °C$) which will minimize interdiffusion of materials in the case of a multilayered thin-film preparation. The film thickness can be controlled by monitoring the amount of charge delivered, whereas the deposition rate can be followed by the variation of the current with time, and schematic arrangement of electrodeposition arrangement is shown in Fig. 4.1. This is based on controlled current or controlled potential electrochemical deposition on an electrode from a solution containing the appropriate species. Therefore, electrodeposition method frequently is used for the deposition of a wide range of compounds including metals, semiconductors, mixed metal oxides and magnetic structures.

Fig. 4.1 Schematic of electrodeposition method

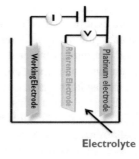

Electrolyte

The electrodeposition technique is particularly well-suited for the deposition of single elements, but it is also possible to carry out simultaneous depositions of several elements and syntheses of well-defined alternating layers of metals and oxides with thicknesses down to a few nanometers. Since, the microstructure of the deposited material can be controlled by adjusting parameters such as the deposition potential or the current density, the composition of the electrolyte or by employing various pulsed deposition techniques, this method can offer a wide range of possibilities for tailor-made syntheses of materials with different structures, morphologies, and properties. Electrochemical method includes cyclic voltammetry, chronoamperometry, and chronopotentiometry to obtain desired products. This is probably the easiest, non-vacuum-based chemical method used for preparing large-area electrode materials. This process is also known as 'electroplating' and is typically restricted to electrically conductive materials.

There are basically two technologies used for plating which are as follows;

(a) Electroplating

In the electroplating process, the substrate is placed in a liquid solution (electrolyte), and an electrical potential is applied between a conducting area on the substrate and a counter electrode in the liquid. A chemical redox process takes place resulting in the formation of a layer of material on the substrate surface on liberation of gas at the counter electrode. Electroplating is a potentially suitable preparation method to obtain low-cost precursor films.

(b) Electroless plating

In the electroless plating process, a more complex chemical solution is used, in which deposition happens spontaneously on any surface which forms a sufficiently high electrochemical potential with the solution. This process is desirable since it does not require any external electrical potential and contact to the substrate during processing. Unfortunately, it is also more difficult to control with regard to film thickness and uniformity. In 1982 and 1984, well-introduced topic of materials science of electrodeposits disclosing how the principles of materials science can be used to explain various structures of electrodeposits and how these structures influence properties are explored in 2011 [65, 66].

Various ferrite thin films synthesized by an electrodeposition method include $NiFe_2O_4$ [67–70], $CuFe_2O_4$ [71], $CoFe_2O_4$ [72–74], Fe_2O_3 [75, 76], and $BiFeO_3$ and mixed bismuth ferrite [77, 78], etc. The electrodeposition method is well suited to make films of metals such as copper, gold, and nickel. The films can be made in any thickness from ~ 1 μm to >100 μm. The surface of the substrate must have an electrically conducting coating before the deposition can be done. The advantage of electrodeposition method is the high-quality film with a very low capital investment.

4.2   Characterization Techniques

The following characterization techniques are frequently used while characterizing
ferrite thin films to identify the phase, morphology, and the composition of the
as-prepared products.

4.2.1   Structural Elucidation

The XRD is a very powerful and suitable technique for characterizing microstructure
of thin films/powders. It is non-destructive and non-contact process that provides
useful information such as structure, phase, grain size, orientation, strain state, etc.,
of material of interest. The basic operation mechanism of XRD can be referred from
various textbooks, e.g., Klug and Alexander, Cullity [79], Tayler [80], Guinier [81],
Barrett, and Massalski [82].

(a)   Operation mechanism

The X-ray scattered by each set of lattice planes at a particular angle, in addition to
the scattered intensity, is a function of the arrangement of atoms in specific crystal
(see Fig. 4.2). The scattering from all the different sets of planes results in a unique
crystal structure pattern of given compound. The X-rays are produced from an anode
(generally copper) in X-ray tube. The copper anode is irradiated with a beam of high-
energy electrons which are accelerated by a high-voltage electric field to a very high
speed. A small window in the X-ray tube allows the X-ray to exit the tube with little
attenuation while maintaining the vacuum seal required for the X-ray tube operation.

Fig. 4.2 Schematic ray
diagram of X-ray diffraction

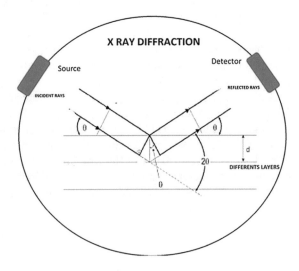

Table 4.1 X-ray diffraction methods

Method	λ	θ
Laue method	Variable	Fixed
Rotating crystal method	Fixed	Variable (in part)
Powder method	Fixed	Variable

The X-rays generally demonstrate 10–0.01 nm wavelength as the lattice spacing of the material under investigation.

Bragg's law describes the diffraction condition from planes with spacing 'd' given by,

$$2d \, \sin \theta \, = \, n\lambda \tag{4.1}$$

where 'd' is the interplanar spacing, 'θ 'is the diffraction angle, 'λ' is the wavelength of X-ray, and 'n' is an order of diffraction.

The way of satisfying Bragg's condition is devised, and this can be done by continuously varying either 'λ 'or 'θ 'during the experiment. The way in which these quantities are varied puts forward three main diffraction methods which are shown in Table 4.1.

In powder method, the crystal to be examined is reduced to a fine powder and placed in a beam of a monochromatic X-rays. Each particle of the powder is the tiny crystal, or assemblage of smaller crystals, oriented at random directions with respect to incident beam. Some of the crystals will be correctly oriented so that their (100) planes, for example, can reflect the incident beam. Other crystals will be correctly oriented for (110) reflections and so on. The result is that every set of lattice planes will be capable of reflection. This is the principle of a powder diffractometer. Ideally, according to Bragg's law, for the particular 'd' value, the constructive interference of X-rays should occur only at particular 'θ 'value, i.e., Bragg's angle, and for all other angles, there should be destructive interference and intensity of diffracted beam will be minimum there.

(b)   Identification of phases

From the 'd' spacing, phases can be identified using the Joint Committee on Powder Diffraction Standards (JCPDS) powder diffraction file, and the reflections can be indexed for concern Miller indices. However, if the size of the diffracting tiny crystal is considerably small, i.e., nm, there is no more complete destructive interference at $\theta \pm d\theta$, which broadens the peak corresponding to diffracted beam in proportion to the size of the tiny crystal. This can be used to calculate the particle size (crystallite size or grain size). The relation for the same is given by Debye–Scherrer and formulated [32] as

$$t = \frac{0.9\lambda}{\beta \cos \theta_B} \tag{4.2}$$

Fig. 4.3 Actual photograph of X-ray diffraction unit (Hanyang University, Korea)

where '*t*' is the crystal size, 'θ_B' is the diffraction angle, 'λ'is the wavelength of X-rays, and 'β'is the line broadening at full width at half maxima (FWHM) epitaxial or polycrystalline (may or may not be oriented) materials are considered as single crystal or powder (crystals or assembly of crystals spread on substrate). Hence, a typical epitaxial or oriented material may not show all corresponding reflections and show only few reflections, for example, say, a *c*-axis-oriented material will show only (*h k l*) for which '*h*' and '*k*' indices are zero and '*l*' is nonzero. However, these hidden peaks can be detected by a small angle X-ray diffraction method. The resultant diffraction lines with obvious peaks together are called an XRD pattern; it can provide information on crystal structure, qualitative chemical composition, and physical properties like strain energy of material under investigation. The basic use of XRD in this book was to determine the actual phase of the as-prepared materials by comparing the obtained XRD pattern to known standard diffraction lines in the JCPDS database. Figure 4.3 shows an X-ray diffractometer (located and at the Hanyang University, Seoul, South Korea, and used for analysis).

4.2.2 Surface Morphology Confirmation

The SEM uses a focused electron probe to extract structural and chemical information point by point from a region of interest in the sample. The high spatial resolution of

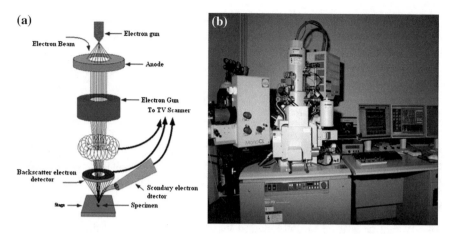

Fig. 4.4 **a** Ray diagram of scanning electron microscopy and **b** a typical FE-SEM facility (Hanyang University, Seoul, South Korea) used for present work

an SEM makes it a powerful tool to characterize a wide range of specimens from the nanometer to micrometer length scales [83].

(a) Basic principle

Accelerated electrons in SEM carry considerable of kinetic energy, and this energy is dissipated as a variety of signals produced by electron–sample interactions when the incident electrons are decelerated in the solid sample. These signals include secondary electrons (that produce SEM images), backscattered electrons (BSE), diffracted backscattered electrons (that are used to determine crystal structures and orientations of minerals), photons (characteristic X-rays that are used for elemental analysis and continuum X-rays), visible light, and heat as shown in Fig. 4.4. Secondary electrons and backscattered electrons are commonly used for imaging samples; secondary electrons are most valuable for showing morphology and topography of sample under consideration, and backscattered electrons are most valuable for illustrating contrasts in composition in multiphase samples [84]. The X-ray generation is produced by inelastic collisions of the incident electrons with electrons in discrete shells of atoms in the sample. As the excited electrons return to lower energy states, they yield X-rays that are of a fixed wavelength. Thus, characteristic X-rays are produced for each element in a mineral that is 'excited' by the electron beam. The SEM analysis is considered to be 'non-destructive,' i.e., X-rays generated by electron interactions do not lead to volume loss of the sample, so it is possible to analyze the same material for several times [85]. The SEM images the surface structure of bulk samples, from the biological, medical, materials sciences, and earth sciences up to magnifications of ~100,000×. The images have a greater depth of field and resolution than optical micrographs which are ideal for rough specimens such as rough and fracture surfaces. A field-emission SEM (FE-SEM) is equipped with a field-emission cathode in the electron gun to provide enhanced resolution and to minimize the charge issues and sample damage. Figure 4.4a, b present ray diagram of scanning electron microscopy and a typical FE-SEM facility (Hanyang University, Seoul, South Korea).

4.2.3 Composition Analysis

Energy-dispersive X-ray (EDX) analysis is a technique to analyze near-surface element and estimate quantitative proportions of elements present at different locations, thus giving an overall elemental mapping of the sample under study.

(a) Basic principle

This technique is used in combination with SEM or FESEM. An electron beam strikes the surface of a conducting sample. The energy of the beam is typically in the range of 10–20 keV. This causes X-ray to be emitted from the material. The energy of the X-ray emitted depends on the material under examination. The X-rays are generated in region about two microns in depth, and thus, EDX is not truly a surface science technique. By moving the electron beam across the material surface, an image of each element in the sample can be obtained. Due to a low X-ray intensity, images usually take a number of hours to acquire. The composition or the amount of particles near and at the surface can be estimated using the EDX, provided they contain some heavy metal ions. The EDX spectra have to be taken by focusing the beam at different regions of same sample to verify uniformity of bimetallic material.

4.2.4 Surface Wettability

The first recognition of wetting phenomenon in scientific research was given by Galileo in 1612. Thomas Young (1805) is considered as the father of scientific research of contact angles and wetting [86]. Wettability is usually quantified in terms of observed contact angles, so from a practical point of view, a simple methodology is needed to account for the heterogeneous rough surface influence on wetting and contact angle measurements. The first attempt at this was made by Wenzel [87]. When a small water drop encounters a solid surface, for example, a raindrop on the hood of a car, a droplet is formed that consists of a sphere of water sectioned by the surface at a discrete, measurable contact angle. The shape of the droplet is not reproducible, and on most surfaces, the contact angle will vary by 20° or more. If a droplet on a surface is allowed to evaporate in a low humidity environment or if water is carefully withdrawn from the droplet using a syringe, the droplet decreases in volume and contact angle maintains the same contact area with the surface until it begins to recede. This is characteristic of the surface of the material under test. If the surface is cooled to below the dew point and water condenses on the droplet or if water is carefully added to the droplet using a syringe, the droplet volume and contact

angle increase and again, the same contact area is maintained until the droplet begins
to advance, which is also characteristic of the surface. So the contact angle gives the
idea wetting behavior of the material surface [88]. If a drop of liquid is placed on a
horizontal solid surface in equilibrium with vapor phase, then the drop spreads on
the solid surface till the three interfacial forces balance each other. Contact angle (θ)
is the angle made by the tangent to the liquid–vapor interface drawn at the contact
line makes an angle with the solid surface, which is the characteristic of the three-
phase system (solid–liquid–vapor). The contact angle, therefore, is a thermodynamic
property.

The three forces act on the contact line due to interfacial tension. At equilibrium
condition, the net force on the contact line must be zero. This is governed by Young's
equation [89].

$$\gamma_{SV} = \gamma_{Sl} + \gamma_{lV} \cos \theta \qquad (4.3)$$

where γ_{SV}, γ_{Sl}, and γ_{lV} are solid–vapor, solid–liquid, and liquid–vapor interfacial
energies, respectively. The 'θ' measurement is crucial in characterizing the surface.
Water contact angle measurement can be performed using contact angle meter. In this
method, the 'θ' can be directly measured using a contact angle meter. A water drop
is kept on the surface, and the image of the drop is projected on the screen with pro-
tractor (refer Fig. 4.5). This gives directly the value of 'θ' after the proper adjustment
of the tangent to the water drop at the point of contact with the solid surface. Nowadays, many industrial applications like lubrication, painting, liquid coating, spray
quenching, soldering, jet printing, etc., are rigorously using wetting and spreading
processes [90]. These applications often employ high-technology materials and surface preparation to control properties related to wettability—adhesion, anticorrosion,
lubrication, friction, wear resistance, biocompatibility, catalysis, antifouling, etc. [91,
92].

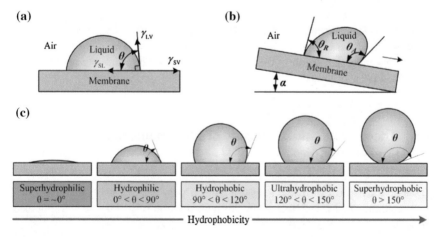

Fig. 4.5 Schematic diagram of contact angle and wettability measurements

4.2.5   Raman Shift Analysis

Raman spectroscopy is a non-destructive spectroscopic technique. Chemists and biologists use Raman spectroscopy to identify structure, functional groups, and to confirm complex biomolecules, such as proteins and deoxyribonucleic acid (DNA). Raman spectroscopy and the infrared absorption spectroscopy are the most widely used techniques that offer information about the structure and properties of molecules by providing vibrational transitions. Raman shift measurement provides key information easily and quickly. This technique can be used for knowing vibrational, rotational, and other low-frequency modes in the system. It relies on inelastic scattering or Raman scattering of monochromatic laser light. The laser light interacts with phonons or other excitations in the systems, resulting in the energy of the laser photons being shifted up or down. The shift in energy gives information about the phonon modes in the system.

(a)   Basic principle

The Raman effect occurs when light impinges upon a molecule that interacts with the electron cloud of the bonds exciting one of the electrons into the virtual state. In spontaneous Raman effect, the molecule will have excited from the ground state to a virtual energy state, and relax into a vibrational excited state, which generates stokes Raman scattering. If the molecule was in already in the elevated vibrational energy state, then Raman scattering is called anti-stokes Raman scattering. The change of amount of polarizability will determine the Raman scattering intensity, whereas the Raman shifts are equal to the vibrational level that is involved.

Raman spectroscopy is commonly used in materials science, since the vibrational information is very specific for the chemical bonds in molecules. Therefore, it provides fingerprint by which the molecules can be identified in the range of $500–2000 \, cm^{-1}$. Raman gas analyzers demonstrate many practical applications. For instance, they are used in medicine for real-time monitoring of anesthetic and respiratory gas mixtures during surgery. In material science and solid-state physics, spontaneous Raman spectroscopy is used to characterize materials, measure temperature, and find the crystallographic orientation of sample. The polarization of the crystal and the polarization of the laser light can be used to find the orientation of the crystal [93].

4.3   Theoretical Understanding of Electrochemical Supercapacitors

The electrochemical measurements are generally carried out using a conventional three-electrode cell filled with the electrolytic solution (Fig. 4.6)—a working electrode (the sample under study), a counter electrode (platinum plate), and a reference

Fig. 4.6 Electrochemical measurement cell conistsing conventional three-electrode system

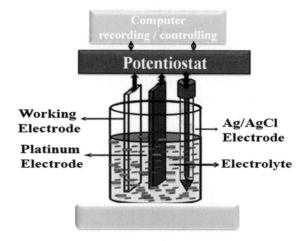

electrode (such as Ag/AgCl/KCl sat.). The exposed surface area must be accurately determined.

4.3.1 Cyclic Voltammetry

Cyclic voltammetry (CV) is electroanalytical technique, also called as linear sweep voltammetry (LSV), probably one of the more versatile techniques available for the electrochemist. The derivation of the various forms of CV can be traced to the initial studies of Matheson and Nicols [94] and Randles [95]. The CV curve is based on the measurement of the current flowing through an electrode dipped in a solution containing electroactive compounds, while a potential scanning is imposed upon it. This electrode is called working electrode and could be made with several materials including metal oxides/chalcogenides. The CV measurement is normally conducted in a three-electrode electrochemical cell containing a working electrode, counter electrode, and reference electrode as shown in Fig. 4.6. The electrolyte in the three-electrode cell is normally an aqueous or non-aqueous liquid solution. During CV measurement, the potential of the working or target electrode in the system is measured against the reference electrode via linear scanning back and forth between the specified upper and lower potential limits. During scanning of the electrode potential (difference between working electrode and reference electrode), the current passing between the working electrode and the counter electrode is recorded. The current passing through the working electrode can be then plotted as a function of electrode potential to yield a CV. The potential range is scanned in one direction, starting at the initial potential and finishing at the final potential.

A more commonly used variation of the technique is CV, in which the direction of the potential is reversed at the end of the first scan. Thus, the waveform is usually of the form of an isosceles triangle. This has the advantage that the product of the electron transfer reaction that occurred in the forward scan can be probed again in the reverse scan. In addition, it is a powerful tool for the determination of formal redox potentials, detection of chemical reactions that proceed or follow the electrochemical reaction and evaluation of electron transfer kinetics.

CV is used to;

(a) reveal surface contamination,
(b) estimate relative surface area and roughness,
(c) evaluate electrolyte leakage at electrode–insulator interfaces,
(d) 'fingerprint' electrochemical reactions for benchmarking and quality control,
(e) estimate deposition potentials at which reduction–oxidation reactions occur, and
(f) determine charge storage capacity.

4.3.2 Galvanostatic Charge–Discharge

Galvanostatic charge–discharge test is electrochemical test in which constant current is applied through an electrolytic test cell. This method is also called chronopotentiometry and gives access to different parameters such as (a) capacitance, (b) resistance, and (c) durability. For three-electrode supercapacitor test cell, the galvanostatic charge–discharge performance is examined by a chronopotentiometry technique on an electrochemistry workstation with an aqueous electrolyte in open circumstances. For a two-electrode configuration (both supercapacitor and lithium battery with an organic electrolyte) in a closely sealed coin cell, a battery tester is commonly used. Galvanostatic charge–discharge tests exhibit electrochemical information on charge–discharge profiles, Columbic efficiency, and long-term cycling properties. Workstation used for measuring electrochemical tests at Hanyang University, Seoul, South Korea, is shown in Fig. 4.7.

This is certainly the most widely used technique in the battery field because it can be extended from a laboratory scale to an industrial level too which is convenient for the systems of high resistance also.

Fig. 4.7 Typical IVIUM: COMPACTSTAT unit (Hanyang University, Seoul, South Korea)

4.3.3 *Electrochemical Impedance Spectroscopy*

Electrochemical impedance spectroscopy (EIS) plays an important role in funda-mental, applied electrochemistry and materials science in the coming years. It is a relatively new and powerful method of characterizing several electrical proper-ties and parameters of materials and their interfaces with electronically conducting electrodes/electrolytes. The concept of electrical impedance was first introduced by Havyside in 1880s. His work extended by A. E. Kennelly and C. P. Steinmetz via use of vector diagrams and complex numbers' representation. In 1899, Warburg deter-mined impedance of diffusional transport of an electroactive species to an electrode surface. Impedance was employed to characterize the dynamics of an electrochemi-cal process in terms of an electrochemical cell's response to an applied potential at different frequencies. The use of EIS in batteries and supercapacitors often provides an estimation of the internal resistance (electrolyte resistance and charge-transfer resistance) at an open-circuit potential or under other conditions.

(a) Operation principle

A more direct technique for studying electrode processes is to measure the change in the electrical impedance of an electrode by EIS. To relate the impedance of the electrode–electrolyte interface for obtaining electrochemical parameters, it is nec-essary to establish an equivalent circuit to represent the dynamic characteristics of respective interface. EIS is a non-stationary technique which is based on the differ-entiation of the reactive phenomena by their relaxation time. The electrochemical system is submitted to a sinusoidal voltage perturbation of low amplitude and vari-able frequency. At each frequency, the various processes evolve with different rates, enabling to distinguish them. The impedance of the diffusion process is often referred to as the 'Warburg impedance.'

In 1903, Kruger [96] realized that the double-layer capacitance had influence on the impedance of the electrode interface and derived an expression for its effect. Much later, the technique was developed and adapted to the study of electrode kinetics [97–101], with emphasis on the charge-transfer process. The charge-transfer step has been considered and analyzed by Randles [98], Grahame [102], Delahay [103], Baticle and Perdu [104, 105], and Sluyters-Rehbach and Sluyters [106], while Gerischer [107, 108] and finally, Michael Gratzel [109] considered coupled homogeneous and heterogeneous chemical reactions while explaining steps in dye-sensitized and per-voskite solar cells. Adsorption processes can be studied by this technique, and the description of the method used to measure adsorption has been given by Laitinen and Randles [110], Llopis et al. [111], Senda and Delahay [112], Sluyters-Rehbach et al. [113], Timmer et al. [114], Barker [109], and Holub et al. [115]. An excellent EIS publication has been written by MacDonald [116] that analyzes the solid/solution interface by equivalent circuitry to calculate the interfacial resistance, capacitance, and inductance values that are reaction mechanisms involved therein.

The impedance data can be represented in two ways:

(a) Nyquist spectrum:- Z_{im} as a function of Z_{re} and
(b) Bode spectrum: log $|Z|$ and phase angle φ as a function of log f.

The impedance measurements are performed over large frequencies ranges, typically from 10 μHz to 1 MHz using amplitude signal voltage in the range of 5–50 mV rms. The amplitude is strongly depending on the studied system. Hence, EIS is a powerful tool, i.e., for investigating the mechanisms of electrochemical reactions, for measuring the dielectric and transport properties of materials, for exploring the properties of porous electrodes, for investigating passive surfaces.

4.4   Conclusions

The bismuth ferrite powders or films prepared by various synthesis methods i.e., hydrothermal, solvothermal, microwave, mechanothermal, seed-hydrothermal, sol-gel, chemical bath deposition, etc., are discussed in detail. Furthermore, characterization techniques like X-ray diffraction, scanning electron microscopy, energy-dispersive X-ray analysis, surface wettability, Raman shift analysis, cyclic voltammetry, galvanostatic charge/discharge, electrochemical impedance spectroscopy, etc., used for estimating various physical and chemical parameters of bismuth ferrites are throughly explored. Brief information about each technique with operation principle and its use for the investigating properties of the metal chelates, under study, has been covered in this chapter.

References

1. J.T. Han, Y.H. Huang, X.J. Wu, C.L. Wu, W. Wei, B. Peng, W. Huang, J.B. Goodenough, Adv. Mater. **18**, 2145–2148 (2006)
2. Y. Li, X.T. Wang, X.Q. Zhang, X. Li, J. Wang, C.W. Wang, Physica E **118**, 113865 (2020)
3. K. Wang, X. Xu, L. Lu, H. Wang, Y. Li, Y. Wu, J. Miao, J.Z. Zhang, Y. Jiang, A.C.S. Appl, Mater. Interfaces **10**, 12698–12707 (2018)
4. Z.H. Jaffari, S.M. Lam, J.-C. Sin, H. Zeng, Mate. Sci. Semi. Proc. **101**, 103–115 (2019)
5. J.A. Darr, J. Zhang, N.M. Makwana, X. Weng, Chem. Rev. **117**, 11125–11238 (2017)
6. X. Wang, W. Mao, Q. Wang, Y. Zhu, Y. Min, J. Zhang, T. Yang, J. Yang, X. Li, W. Huang, RSC Adv. **7**, 10064–10069 (2017)
7. K. Suzuki, Y. Tokudome, H. Tsuda, M. Takahashi, J. Appl. Cryst. **49**, 168–174 (2016)
8. S. Li, Y.H. Lin, B.P. Zhang, Y. Wang, C.W. Nan, J. Phys. Chem. C **114**, 2903–2908 (2010)
9. Z.T. Hu, S.K. Lua, X. Yan, T.T. Lim, RSC Adv. **5**, 86891–86900 (2015)
10. Q. Zhang, D. Sando, V. Nagarajan, J. Mater. Chem. C **4**, 4092–4124 (2016)
11. L. Fei, J. Yuan, Y. Hu, C. Wu, J. Wang, Y. Wang, Cryst. Growth Des. **11**, 1049–1053 (2011)
12. M.A. Iqbal, A. Tariq, A. Zaheer, S. Gul, S.I. Ali, M.Z. Iqbal, D. Akinwande, S. Rizwan, ACS Omega **4**, 20530–20539 (2019)
13. S. Mondal, K. Dutta, S. Dutta, D. Jana, A.G. Kelly, S. De, A.C.S. Appl, Nano Mater. **1**, 625–631 (2018)

14. M.S. Angotzi, A. Musinu, V. Mameli, A. Ardu, C. Cara, D. Niznansky, H.L. Xin, C. Cannas, ACS Nano **11**, 7889–7900 (2017)
15. S.S.M. Bhat, H.W. Jang, Chemsuschem **10**, 3001–3018 (2017)
16. X. Sun, Z. Liu, H. Yu, Z. Zheng, D. Zeng, Mater. Lett. **219**, 225–228 (2018)
17. K. Chybczynska, M. Blaszyk, B. Hilczer, T. Lucinski, M. Matczak, B. Andrzejewski, Mate. Res. Bull. **86**, 178–185 (2017)
18. C. Ponzoni, M. Cannio, D.N. Boccaccini, C.R.H. Bahl, K. Agersted, C. Leonelli, Mater. Chem. Phys. **162**, 69–75 (2015)
19. A. Perejona, E.G. Gonzaleza, P.E. Sanchez-Jimeneza, A.R. Westc, L.A. Perez-Maqueda, J. Euro, Ceram. Soc. **39**, 330–339 (2019)
20. J. Rout, R.N.P. Choudhary, Phys. Lett. A **380**, 288–292 (2016)
21. A.R. Goldmana, J.L. Fredricksa, L.A. Estrof, J. Crystal Growth **468**, 104–109 (2017)
22. S. Bharathkumar, M. Sakar, S. Balakumar, J. Phys. Chem. C **120**, 18811–18821 (2016)
23. A. Soam, R. Kumar, C. Mahender, M. Singh, D. Thatoi, R.O. Dusane, J. Alloy. Comp. **813**, 152145 (2020)
24. R.A. Golda, A. Marikani, E.J. Alex, Ceram. Int. **46**, 1962–1973 (2020)
25. X. Yuan, L. Shi, J. Zhao, S. Zhou, X. Miao, J. Guo, A.C.S. Appl, Nano Mater. **2**, 1995–2004 (2019)
26. A. Castro, M.A. Martins, L.P. Ferreira, M. Godinho, P.M. Vilarinho, P. Ferreira, J. Mater. Chem. C **7**, 7788–7797 (2019)
27. S.D. Waghmare, V.V. Jadhav, S.F. Shaikh, R.. Mane, J.H. Rheee, C. O'Dwyer, Sens. Actuators A **271**, 37–43 (2018)
28. T.J. Park, G.C. Papaefthymiou, A.J. Viescas, A.R. Moodenbaugh, S.S. Wong, Nano Lett. **7**, 766–772 (2007)
29. R. Das, G.G. Khan, S. Varma, G.D. Mukherjee, K. Mandal, J. Phys. Chem. C **117**, 20209–20216 (2013)
30. L.L. Hench, J.K. West, Chem. Rev. **90**, 33–72 (1990)
31. C. Lausecker, B. Salem, X. Baillin, V. Consonni, J. Phys. Chem. C **123**, 29476–29483 (2019)
32. V.V. Jadhav, S.A. Patil, D.V. Shinde, S.D. Waghmare, M.K. Zate, R.S. Mane, S.H. Han, Sens. Actuators, B **188**, 669–674 (2013)
33. S.Y. Lim, S. Park, S.W. Im, H. Ha, H. Seo, K.T. Nam, ACS Catal. **10**, 235–244 (2020)
34. D.K. Pawar, S.M. Pawar, P.S. Patil, S.S. Kolekar, J. Alloy. Comp. **509**, 3587–3591 (2011)
35. J.L. Gunjakar, A.M. More, V.R. Shinde, C.D. Lokhande, J. Alloy. Comp. **465**, 468–473 (2008)
36. V.S. Kumbhar, A.D. Jagadale, N.M. Shinde, C.D. Lokhande, App. Surf. Sci. **259**, 39–43 (2012)
37. C.M. Raghavan, J.W. Kim, S.S. Kim, Ceram. Int. **39**, 3563–3568 (2013)
38. S.R. Chitra, S. Sendhilnathan, Int. J. Appl. Ceram. Technol. **12**, 643–651 (2015)
39. C. Aoopngan, J. Nonkumwong, S. Phumying, W. Promjantuek, S. Maensiri, P. Noisa, S. Pinitsoontorn, S. Ananta, L. Srisombat, A.C.S. Appl, Nano Mater. **2**, 5329–5341 (2019)
40. U.A. Agu, S.N. Mendieta, M.V. Gerbaldo, M.E. Crivello, S.G. Casuscelli, Ind. Eng. Chem. Res. (2019). https://doi.org/10.1021/acs.iecr.9b04042
41. S.L. Hua, J. Liub, H.Y. Yuc, Z.W. Liuc, J. Mag. Magnet. Mater. **473**, 79–84 (2019)
42. P. Sharma, P.K. Diwan, O.P. Pandey, Mat. Chem. Phy. **233**, 171–179 (2019)
43. R. Das, R. Karna, Y.C. Lai, F.C. Chou, Cryst. Growth Des. **16**, 499–503 (2016)
44. E.S. Kim, N. Nishimura, G. Magesh, Y. Kim, J.W. Jang, H. Jun, J. Kubota, K. Domen, J.S. Lee, J. Am. Chem. Soc. **135**, 5375–5383 (2013)
45. E. Carpenter, V.G. Harris, J. Appl. Phys. **91**, 7589–7591 (2002)
46. A. Safari, Kh Gheisari, M. Farbod, J. Mag. Magnet. Mater. **488**, 165369 (2019)
47. M.L. Boulos, **1**, 33–40 (1992)
48. S. Son, R. Swaminathan, M.E. McHenry, J. App. Phy. **93**, 7495–7497 (2003)
49. S.P. Dalawai, S. Kumar, M.A.S. Aly, M.Z.H. Khan, R. Xing, P.N. Vasambekar, S. Liu, J. Mat. Sci. Mat. Elect. **30**, 7752–7779 (2019)
50. R.A. Henning, P. Uredat, C. Simon, A. Bloesser, P. Cop, M.T. Elm, R. Marschall, J. Phys. Chem. C **123**, 18240–18247 (2019)

51. M.V. Rastei, F. Gelle, G. Schmerber, A. Quattropani, T. Fix, A. Dinia, A. Slaoui, S. Colis, A.C.S. Appl, Energy Mater. **2**, 8550–8559 (2019)
52. G. Bulai, V. Trandafir, S.A. Irimiciuc, L. Ursu, C. Focsa, S. Gurlui, Ceram. Int. **45**, 20165–20171 (2019)
53. F. Eskandari, P. Kameli, H. Salamati, Appl. Surf. Sci. **466**, 215–223216 (2019)
54. M.K. Shamim, S. Sharma, R.J. Choudhary, J. Alloys Compd. **794**, 534–541 (2019)
55. P.C. Dorsey, S.E. Bushnell, R.G. Seed, C. Vittoria, J. Appl. Phys. **74**, 1242–1246 (1993)
56. P.C. Dorsey, S.E. Bushnell, R.G. Seed, C. Vittoria, I.E.E.E. Trans, Mag. **29**, 3069–3071 (1993)
57. B.M. Simion, R. Ramesh, V.G. Keramidas, G. Thomas, E. Marinero, R.L. Pfeffer, J. Appl. Phys. **76**, 6287–6289 (1994)
58. D. Hong, Y. Yamada, M. Sheehan, S. Shikano, C.H. Kuo, M. Tian, C.K. Tsung, S. Fukuzumi, ACS Sustainable Chem. Eng. **2**, 2588–2594 (2014)
59. S. Hussain, S. Hussain, A. Waleed, M.M. Tavakoli, S. Yang, M.K. Rauf, Z. Fan, M.A. Nadeem, J. Phys. Chem. C **121**, 18360–18368 (2017)
60. E. Freeman, S. Kumar, V. Celorrio, M.S. Park, J.H. Kim, D.J. Fermin, S. Eslava, Sustainable Energy Fuels (2020)
61. L.X. Phua, F. Xu, Y. G. Ma, C.K. Ong, Thin Solid Films **517**, 5858–5861 (2009)
62. H. Arabi, N.K. Moghadam, J. Mag. Magnet. Mater. **335**, 144–148 (2013)
63. J. Leng, Z. Wang, J. Wang, H.H. Wu, G. Yan, X. Li, H. Guo, Y. Liu, Q. Zhang, Z. Guo, Chem. Soc. Rev. **48**, 3015–3072 (2019)
64. Y. Li, J. Shen, Y. Hu, S. Qiu, G. Min, Z. Song, Z. Sun, C. Li, Ind. Eng. Chem. Res. **54**, 9750–9757 (2015)
65. R. Weil, Plat. Surf. Finish. **69**, 46 (1982)
66. J. Duay, E. Gillette, J. Hu, S.B. Lee, Phys. Chem. Chem. Phys. **15**, 7976–7993 (2013)
67. C.D. Lokhande, S.S. Kulkarni, R.S. Mane, O.S. Joo, S.H. Han, Ceram. Int. **37**, 3357–3360 (2011)
68. S.D. Sartale, C.D. Lokhande, Ind. J. Eng. Mater. Sci. **7**, 404–410 (2000)
69. S.D. Sartale, C.D. Lokhande, M. Gersig, V. Ganesan, J. Phys. Condens. Mater. **15**, 773–778 (2004)
70. J.L. Gunjakar, A.M. More, V.R. Shinde, C.D. Lokhande, J. Alloy. Compd. **465**, 468–473 (2008)
71. U. Kang, S.K. Choi, D.J. Ham, S.M. Ji, W. Choi, D.S. Han, A.A. Wahab, H. Park, Energy Environ. Sci. **8**, 2638–2643 (2015)
72. H. He, N. Qian, N. Wang, Cryst. Eng. Comm. **17**, 1667–1672 (2015)
73. Z. He, J.A. Koza, G. Mu, A.S. Miller, E.W. Bohannan, J.A. Switzer, Chem. Mater. **25**, 223–232 (2013)
74. Z. Jia, L. Yang, Q. Wang, J. Liu, M. Ye, R. Zhu, Mater. Chem. Phys. **145**, 116–124 (2014)
75. B.J. Lokhande, R.C. Ambare, S.R. Bharadwaj, Measurement **47**, 427–432 (2014)
76. J.H. Lim, S.G. Min, L. Malkinski, J.B. Wiley, Nanoscale **6**, 5289–5295 (2014)
77. C.D. Lokhande, T.P. Gujar, V.R. Shinde, R.S. Mane, S.H. Han, Electro. Commun. **9**, 1805–1809 (2007)
78. V.V. Jadhav, M.K. Zate, S. Liu, M. Naushad, R.S. Mane, K.N. Hui, S.H. Han, Appl Nanosci. **6**, 511–519 (2016)
79. G.R. Chatwal, S.K. Anand, *Instrumental Methods of Chemical Analysis* (Himalaya Publishing House, 2006)
80. M.J. Buerger, *X-ray Crystallography* (Wiley, New York, 1942)
81. H.P. Klug, L.E. Alexander, *X-ray Diffraction Procedures* (Wiley, New York, 1954)
82. B.D. Cullity, *Elements of X-ray Diffraction* (Addison Wesley, Massachusetts, 1956)
83. J. Goldstein, *Scanning Electron Microscopy and X-ray Microanalysis* (Kluwer Academic/Plenum Pulbishers, New York, 2003)
84. L. Reimer, *Scanning Electron Microscopy: Physics of Image Formation and Microanalysis* (Springer, 1998)
85. R.F. Egerton, *Physical Principles of Electron Microscopy: An introduction to TEM, SEM, and AEM* (Springer, 2005)

86. J. Robert, J. Good, Adh. Sci. Techn. **6**(12), 1269 (1992)
87. R.N. Wenzel, Ind. Eng. Chem. **28**, 988 (1936)
88. P.G. De Gennes, Rev. Mod. Phys. **830** (1985)
89. A.W. Anderson, (John Wiley, New York, 1982), 338
90. K. Narayan Prabhu, P. Fernades, G. Kumar, Mater. Design **30**, 297 (2009)
91. V. Roucoules, F. Gaillard, T. Mathia, P. Lanteri, Adv. Colloid Interf. Sci. **97**, 177 (2002)
92. A. Borruto, G. Crivellone, F. Marani, Wear **222**, 57 (1998)
93. G. Gouadec, P. Colomban, Progress in crystal growth and characterization of materials. Elsevier **53**, 1 (2007)
94. L.A. Matheson, N. Nichols, J. Electrochem. Soc. **73**, 193 (1938)
95. J.E.B. Randles, Trans. Faraday Soc. **44**, 327 (1948)
96. F. Kruger, Z. Phys, Chem. **45**, 1 (1903)
97. P. Dolin, B. V. Ershler, Acta Physicochim. U.S.S.R. **13**, 747 (1940)
98. J.E.B. Randles, Disc. Faraday Soc. **1**, 11 (1947)
99. B.V. Ershler, Disc. Faraday Soc **1**, 269 (1947)
100. E. Sabatani, I. Rubinstein, J. Phys. Chem. **91**(27), 6663 (1987)
101. R.S. Nicholson, Anal. Chem. **37**(11), 1351 (1965)
102. D.C. Grahame, J. Electrochem. Soc. **99**, 370C (1952)
103. P. Delahay, Record Chem. Prog. **19**, 83 (1958)
104. A.M. Baticle, F. Perdu, J. Electroanal. Chem. **12**, 15 (1966)
105. A.M. Baticle, F. Perdu, J. Electroanal. Chem. **13**, 364 (1967)
106. M. Sluyters-Rehbach, J.H. Sluyters, Rec. Trav. Chim. **82**, 525 (1963)
107. H. Gerischer, Z. Phys, Chem. **198**, 286 (1951)
108. H. Gerischer, Z. Phys, Chem. **201**, 55 (1952)
109. P. Bonhote, J. Moser, R. Humphry-Baker, N. Vlachopoulos, S.M. Zakeeruddin, L. Walder, M. Gratzel, J. Am. Chem. Soc. **121**(6), 1324 (1999)
110. H.A. Laitinen, J.E.B. Randles, Trans. Faraday Soc. **51**, 54 (1955)
111. J. Llopis, J. Fernandez-Biarge, M. Perez-Fernandez, Electrochim. Acta **1**, 130 (1959)
112. M. Senda, P. Delahay, J. Phys. Chem. **65**, 1580 (1961)
113. M. Sluyters-Rehbach, B. Timmer, J.H. Sluyters, J. Electroanal. Chem. **15**, 151 (1967)
114. B. Timmer, M. Sluyters-Rehbach, J.H. Sluyters, J. Electroanal. Chem. **15**, 343 (1967)
115. K. Holub, G. Tesari, P. Delahay, J. Phys. Chem. **71**, 2612 (1967)
116. J.R. MacDonald, Annal. Biomed. Eng. **20**, 289 (1992)

Chapter 5
Electrochemical Supercapacitors
of Bismuth Ferrites

5.1 Introduction

Nowadays, due to a fast-growing market for portable electronic devices such as mobile phones, laptops, notebook computers, and uninterruptible power supply systems (UPS) and the development of hybrid electric vehicles, there has been an ever increasing and urgent demand for environmentally friendly high-power energy storage resources. A supercapacitor is an assuring energy storage device, which can act as a gap bridging function between the batteries and conventional capacitors [1–3]. Supercapacitive performance of a material either in thin-film form or in pallet form as electrode material can be assayed by cyclic voltammetry (CV) and galvanostatic charge–discharge (GCD) measurements. An electrode is judged by its specific capacitance value and the number of charge–discharge cycles it withstands, maintaining the constancy of capacitance. This brings forth the search for a wide variety of materials. In general, transition metal oxides including RuO_2, MnO_2, NiO, CO_3O_4, SnO_2, ZnO, TiO_2, V_2O_5, CuO, Fe_2O_3, WO_3 [4–17], etc., have demonstrated a high specific capacitance due to their redox behavior.

Ferrites, one of the special class of the materials, have attracted great attention in energy storage devices due to their cost-effectiveness, abundance, and eco-friendly nature which is highly demanded in material-based electronic industries. Most fascinating applications of ferrite materials are antenna rods, transformer cores, recording heads, loading coils, memory, and microwave devices, etc. [18–20]. Today's electronic industry tends constantly toward tininess of devices and the development of new materials to replace existing one when their inherent technological limits are being reached. In the field of smart materials, a research topic based on multiferroic materials is at the top [21, 22]. BFO (bismuth ferrite), one of the very few multiferroics that approves a simultaneous coexistence of ferroelectric and antiferromagnetic order parameters in the perovskite structure, has been attracted much attention. Potential applications of BFO in the memory devices, sensors, spintronic devices, satellite communications, optical filters, and smart devices, etc., are limited on account of its low insulation resistance caused by the reduction of Fe^{3+} species

© The Authors 2020

V. V. Jadhav et al., *Bismuth-Ferrite-Based Electrochemical Supercapacitors*,
SpringerBriefs in Materials, https://doi.org/10.1007/978-3-030-16718-9_5

to Fe^{2+} and oxygen vacancies for charge compensation [23–25]. The preparation of phase pure BFO without impurities is a critical task due to its narrow temperature stability range [26]. Using conventional solid-state reaction routes, preparation of the BFO material endows following difficulties;

(a) It is difficult to get the pure phase BFO, i.e., $BiFeO_3$, as while forming BFO, other thermodynamically more stable phases such as $Bi_2Fe_4O_9$, $Bi_{46}Fe_2O_{72}$, $Bi_{25}FeO_{40}$, $Bi_{24}Fe_2O_{39}$, and $Bi_{36}Fe_2O_{57}$ are dominant.
(b) Generally, the platinum crucible is needed for its preparation, which can be used only for three or four times which eventually increases preparation cost.
(c) During the deposition, the substrate must be placed at the elevated temperatures, and therefore, this causes restrictions on the selection of a substrate material.

Single-phase BFO powder can be prepared by oxide mixing technique followed by leaching with dilute nitric acid to eliminate unreacted impurity phases [27]. Palkar and Pinto [28] synthesized highly resistive thin film of phase pure BFO using pulsed laser deposition method. Epitaxial BFO, i.e., $Bi_3Fe_5O_{12}$ is prepared by a reactive ion beam sputtering technique by Adachi et al. [29]. The BFO nanostructures including nanofibers [30], nanowires [31, 32], nanorods [33], nanoparticles [34–36], microcubes, and nanoplates [37] are already been obtained which are expected to offer a better efficiency owing to their large surface area for better specific capacitance. Electrodeposition is a powerful and interesting methodology that can be applied in numerous fields for synthesizing thin metal oxide/chalcogenide/hydroxide/layered double hydroxide/halide thin films or coatings. It is also used for coating a few metallic layers too. Films can be synthesized at a low or room temperature because of high-energy density accumulated in solution near the electrode surface. The advantages of electrodeposition method compared with other methods include; (a) low-cost for raw materials and equipments, (b) capability of controlling composition and morphology by varying electrochemical parameters, and (c) the ability to deposit films on a complex surface. This is probably the easiest, non-vacuum, and suitable method used to prepare electrodes of large area. It has been used for the preparation of thin and thick films of iron group metal oxide at relatively low temperatures [38–40]. Lokhande et al., [41] have obtained specific capacitance of 81 F g^{-1} of BFO films prepared by electrodeposition method. Dutta et al., [42] have obtained a high value of specific capacitance of 450 F g^{-1} for BFO nanorods prepared by a wet chemical template method.

5.2 Bismuth Ferrites

Multiferroics is a class of multifunctional materials which simultaneously demonstrate ferroelectricity, ferromagnetism, ferroelasticity, etc., effects as a result of the coupled electric, magnetic, mechanical, and structural properties [43–45]. In recent years, several efforts have been made in designing and synthesizing this type of multifunctional materials as they exhibit exotic properties at the nanoscale [46–49].

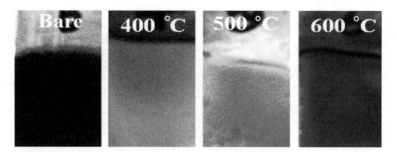

Fig. 5.1 Digital photo images of mixed-phase bismuth ferrites for different annealing temperatures

Bismuth ferrite ($BiFeO_3$, BFO), a multiferroic material with perovskite structure was first recognized in 1960 [50]. But, they didn't obtained as a single crystal. Polycrystalline ceramics are not usable for practical applications owing to their high conductivity [50, 51]. In 1967, Achenbach et al. [52] have succeeded in preparing single-phase polycrystalline $BiFeO_3$ by removing the unwanted phases using HNO_3. In 1990, Kubel and Schmid [53] have carried out precise X-ray diffraction study on monodomain single crystal of $BiFeO_3$. The discovery on $BiFeO_3$ thin films of large remnant polarization showed improvement fifteen times larger than that of previously obtained for the bulk, together with strong ferromagnetism [54]. Many other studies are also carried out on the bulk, thin-film, and nanostructured $BiFeO_3$ since then [55–64].

McDonnell et al. [65] have synthesized mixed BFO on stainless steel using electrodeposition technique at room temperature and annealed at 400, 500, and 600 °C for phase confirmation as shown in Fig. 5.1.

5.2.1 Crystal Structure

The BFO is an inorganic chemical compound and an unusual compound of bismuth, iron, and oxygen. The room temperature phase of $BiFeO_3$ has a rhombohedrally distorted perovskite structure with *R3c* space group [48, 60] as shown in Fig. 5.2.

The unit cell of $BiFeO_3$ can be described by pseudocubic, rhombohedral, or hexagonal settings [66, 67]. It is one of the most promising lead-free piezoelectric materials, which exhibits multiferroic properties at room temperature. The BFO demonstrates 2.2–2.8 eV band gap energy [68–70] whose lattice constants are $a_{pc} = 3.965$ Å and $\alpha_{pc} = 89.35°$ for the pseudocubic unit cell containing one formula unit, $a_{rh} = 5.6343$ Å and $\alpha_{rh} = 59.348$ in the rhombohedral unit cell containing two formula units, $a_{hex} = 5.578$ Å and $c_{hex} = 13.868$ Å, and for the hexagonal unit cell containing six formula units [65]. The temperature has an important effect on the crystal structure and lattice spacing of bismuth ferrite. The atomic coordinates in the unit cell of $BiFeO_3$ are almost unchanged between 5 K and 300 K operating temperatures. The z atomic coordinate values of Fe and O ions, which determine the $BiFeO_3$ unit

Fig. 5.2 Schematic view of the *R3c* structure built up from two cubic perovskite unit cells

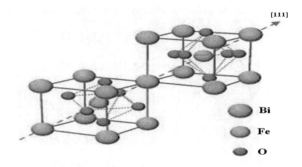

cell polarization, increase considerably above 300 K [71]. At room temperature, the structural symmetry is that of a rhombohedral distorted ABO_3 perovskite where A and B are two metal cation and B is anion. It undergoes a structural phase transition near 1098 K to a β-phase that is orthorhombic, and above 1204 K the γ-phase structure is cubic [72]. It has a ferroelectric Curie temperature T_c of 850 °C and an antiferromagnetic Neel temperature of 370 °C [73, 74].

5.2.2   Structure confirmation

The X-ray diffraction technique is frequently used investigate the structural identification, lattice parameters determination, phases, and crystallite orientation of the material. Figure 5.3 presents XRD patterns of the BFO electrodes annealed at 400, 500, and 600 °C. Mixed-phases $Bi_2F_4O_9$, $BiFeO_3$, and $Bi_{25}FeO_{40}$ BFO electrodes are examined at 600 °C.

Fig. 5.3 The XRD patterns of BFO annealed at 400, 500, and 600 °C

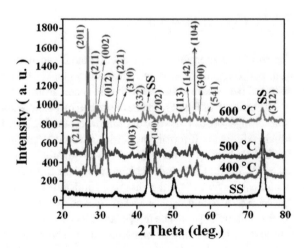

5.2.3 Surface Appearance and Compositional Analysis

The FE-SEM images of BFO electrodes annealed at 400, 500, and 600 °C are as shown in Fig. 5.4. Here, all images are closely matched to one another. There is no substantial difference in the surface appearance. All surfaces are smooth-polished and compact, which would be one of the drawbacks for using them in ES application as ions of electrolyte would face moderate hindrance to reach at BFO matrix. It was concluded that by changing the electrodeposition condition and composition of solution, the structural morphology can be greatly changed. From the FE-SEM imaging, the mixed phases of BFO electrode annealed at different temperature revealed irregular flake-like structure. An EDX pattern of the BFO is shown in Fig. 5.3. The as-obtained Bi, Fe, and O elemental percentages are well supporting to BFO formation (1:2:4). Considerable voids are noted. Surface of the BFO annealed at 500 °C is compact and well-polished, but voids are decreased considerably. From the BFO surface annealed at 600 °C, the presence of sharp-edged crystallites is confirmed.

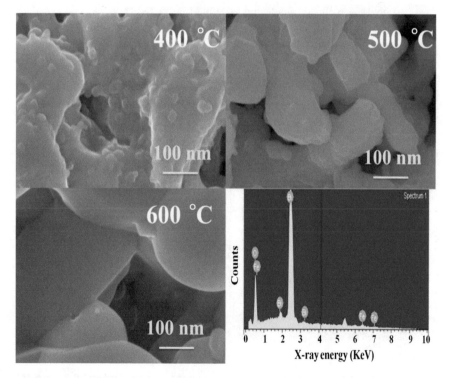

Fig. 5.4 FE-SEM surface images of BFO films annealed at 400, 500, and 600 °C temperatures

5.2.4   Surface Wettability Study

The wetting behavior can be studied with the help of contact angles measured over various positions of BFO surface. The photographs of contact angles are shown in Fig. 5.5. The contact angle factor is known to effect on the surface energy followed by ES performance which is function of pore size, morphology, crystal structure, crystal type, and moreover, internal and external strains. The contact angle measured on BFO surface decreases systematically as the annealing temperature is increased. It is believed that the hydrophilic surface reduces the diffusion resistance of the electrolyte for better interfacial kinetics.

The contact angle measured on BFO surface decreases systematically as the annealing temperature is increased. It is believed that the hydrophilic surface reduces the diffusion resistance of the electrolyte for better interfacial kinetics. If the surface wettability is high, contact angle will be small, and the surface is hydrophilic. On the contrary, if the wettability is low, h will be large, and the surface is hydrophobic. Due to change in surface morphology, the contact angle was decreased from 86.21° to 58.48°. With the increasing annealing temperature, water contact angle

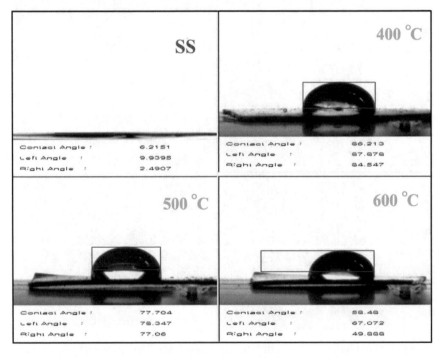

Fig. 5.5   Contact angle measurements on SS and BFO annealed at; **a** 400 °C, **b** 500 °C, and **c** 600 °C surfaces

value was decreased. Lower contact could be beneficial for achieving enhanced specific capacitance [75–77]. At 600 °C, BFO electrodes are superhydrophilic with the lower contact angle being beneficial for achieving enhanced specific capacitance.

5.2.5 Electrochemical Performance Study

The synthesized nanoflake electrode exhibited high specific capacitance of 72.2 F g^{-1} at a current density 1 A g^{-1} in 2 M NaOH electrolyte as shown in Fig. 5.6. The specific capacitance at various current densities is calculated based on their galvanostatic charge and discharge. This excellent performance attributed the high surface area of nanoflakes and fast ion transfer by enhancing faradic redox reaction. At lower scan rates fully, utilization of the BFO is possible thereby at lower scan rate inner and outer sites are active while at higher scan rates only surface sites are generally active which results lesser SC performance. The capacitance retention is about 37% up to 20 A g^{-1} which is having excellent cycling stability with 82.8% retention after 1500 cycles.

The galvanostatic charge/discharge profiles of BFO in 2 M NaOH electrolyte are shown in Fig. 5.6b. The charge–discharge current rate is 1 mA g^{-1}, and the operational

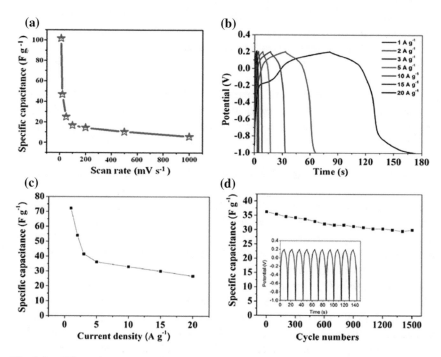

Fig. 5.6 **a** Effect of scan rate on SC value of BFO electrode. **b** Galvanostatic charge–discharge curves of the BFO at different current densities. **c** Specific capacitance under different current densities. **d** Capacitance retention of BFO for 1500 cycles

potential range is between -1.0 and $+0.2$ V. The nonlinear discharging nature is in accordance with the capacitive performance which is due to a combined effect of electric double layer and pseudocapacitance share. Ying Lin et al. [78] reported $Bi_{0.9}La_{0.1}FeO_3$ (BLFO) porous microspheres with high surface area having largest specific capacitance of 561.48 F g^{-1} at a scan rate of 2 mV s^{-1}. This shows that the mixed BFO of this film electrode could be a potential candidate for supercapacitor electrode.

5.3 Mixed Bismuth Ferrites

The term multiferroic has been used for the first time by Schmid in 1994 [79]. Discovery of multiferroics materials has generated a great deal of interests because of their potential applications in electronic devices [80, 81] on account of combined magnetic and electric properties. Naturally occurring multiferroics include Cr_2O_3, Ti_2O_3, $GaFeO_3$, $PbFe_{0.5}Nb_{0.5}O_3$, and numerous others [82–85]. Particularly, they are used in data storage processing [86, 87] and direct conversion of heat to electricity [88] as they demonstrate combined ferroelectricity, ferromagnetism, or antiferromagnetic properties. The $CoFe_2O_4$–$BiFeO_3$ (CFO–BFO) composite system has generated serious concern due to the magnetoelastic properties of CFO [89] and the combination of ferroelectricity and antiferromagnetism of BFO [90]. The combination of BFO and CFO demonstrates different and special crystal structures, for example, a perovskite configuration for the bismuth and a spinel configuration for the cobalt which are naturally immiscible.

In Massachusetts Institute of Technology (MIT), materials science and engineering department has developed new method for making multiferroic complex metal oxide thin films by a pulsed laser deposition while controlling their magnetic properties. Aimon et al. reported nanopillars of CFO in a matrix of BFO on a strontium titanate ($SrTiO_3$) substrate [91]. The coupling between the magnetic and the electrical properties of complex metal oxides like CFO/BFO nanocomposite one can use them in electrically switchable magnetic data storage devices. Several groups have already tried to fabricate composite ferroelectric/ferromagnetic nanostructures [92]. The challenge now is to develop comparable materials in thin-film form for integrated technologies [93]. Attention has mainly concentrated in composite nanostructures of perovskite and spinel structure oxides like $BiFeO_3$ or $BaTiO_3$ and $CoFe_2O_4$ [94–96]. Nanocomposites are another wide scope research area consisting of multiferroic materials [97, 98]. The first multiferroic composite was created from ferroelectric $BaTiO_3$ and ferromagnetic $CoFe_2O_4$ by unidirectional solidification in a eutectic composite [99]. Recently, many multiferroic nanocomposite materials like $PbTiO_3$–$CoFe_2O_4$, [100] and $BiFeO_3$–$NiFe_2O_4$ [101] have prepared. A special focus has been set on the growth and characterization of multiferroic heterostructures, where nanopillars of one material are embedded into the matrix of the other [102]. Also CFO–BFO was codeposited via physical vapor deposition at high temperatures on a $SrTiO_3$ [103]. The BFO–CFO nanocomposite structure was formed using anodic aluminum oxide [104]. The BFO has been acknowledged to be a promising alternative of lead-based dielectrics/ferroelectrics with a large spontaneous polarization (Ps) of ~100 μC cm^{-2} and a high Curie temperatures (T_c) of 830 °C [105]. But, pure BFO exhibits a large remnant polarization (Pr) due to strong ferroelectric (FE) hysteresis, restricting its usage for energy storage [105]. Modified BFO-based dielectrics such as BFO–BTO, BFO–$(Bi_{1/2}Na_{1/2})TiO_3$, and BFO–$Pb(Zr,Ti)O_3$ have been reported to show relaxor ferroelectric (RFE)-like features [106–108]. Some preliminary work also found RFE properties with potentially good energy performance in BFO-$SrTiO_3$ systems [109, 110]. However, the underlying mechanisms for the emergence of RFE features in BFO-based dielectrics and a feasible approach to design high-energy density BFO-based RFEs remain undiscovered. Lin et al. [111] reported a high-performance film dielectric for capacitive energy storage has been a great challenge for modern electrical devices. Here, giant energy densities of ~70 J cm^{-3}, together with high efficiency as well as excellent cycling and thermal stability, can be achieved in lead-free bismuth ferrite strontium titanate solid solution films through domain engineering. This corresponds with the first-principle simulations by Xu et al. [112], proving the huge potential of BFO-based materials as high-energy density dielectrics. As the simulations predicted even higher-energy density (100–150 J cm^{-3}), there are still more rooms for further improvement in the $(BiFeO_3)_{1-x}$–$(SrTiO_3)_x$ system. Moreover, the incorporation of strontium titanate transforms the ferroelectric microdomains of bismuth ferrite into highly dynamic polar nanoregions, resulting in a ferroelectric to relax or ferroelectric transition with concurrently improved energy density and efficiency. On the other hand, the poor conductivity of ferrite-based materials often adversely affects their rate capability and supercapacitive performance [113]. Construction of high-performing supercapacitors using graphene as active species can be attractive strategy due to fact that the graphene provides faster electron transfer pathways with improved chemical stability [113–119]. Ghosh et al. [120] have reported a nanocomposite, composed of a $BiFeO_3$ nanowire and reduced graphene oxide (BFO–RGO), as an electrode material with highest value of specific capacitance 928.43 F g^{-1} at 5A g^{-1} current density. The Cu-doped $BiFeO_3$ was studied by Maensiri et al. with 568.13 F g^{-1} specific capacitance contribution [121]. In addition, no sudden potential drop in the charge–discharge curve of the nanocomposite is evidenced, suggesting the ohmic resistance for the bulk of the nanocomposite must be too low [122].

5.4 Hybrid Supercapacitor Devices of Bismuth Ferrites

It is a well-known fact that the three-electrode electrochemical measurement system is employed to determine the electrochemical behavior of the electrode materials for supercapacitors, but a two-electrode test cell measurement provides more reliable data for practical applications because it mimics the cell configuration of commercial supercapacitors [114, 123]. However, it is also a known fact that the specific capacitance of electrode material when obtained from the two-electrode system is usually lower than that obtained from a three-electrode system [123]. Some hybrid materials of BFO envisaged previously in supercapacitor application are summarized in Table 5.1.

BiFeO$_3$-graphene nanocomposite is synthesized using sol–gel as efficient electrode for supercapacitor application [129]. It gives 64 F g^{-1} at the scan rate of 20 mV s^{-1} in 1 M Na$_2$SO$_4$ electrolyte. Sarkar et al. [127] have synthesized three-dimensional BiFeO$_3$ anchored on anodized TiO$_2$ nanotubes by using chemical bath deposition technique for pseudocapacitors and solar energy conversion application as shown in Fig. 5.7. In this case, BFO nanoparticles having diameter in the range of 2–8 nm considerably enhance the specific surface area anchored TiO$_2$ nanotubes acts as the pathway for electron transportation toward the current collector, making the excellent device for pseudocapacitor electrode. This type of heterostructures exhibits the specific capacitance of about 440 F g^{-1} at a current density of 1.1 A g^{-1} with ~80% capacity retention at current density of 2.5 A g^{-1}. It also exhibits high energy and power density of 46.5 W h kg^{-1} and power density of 1.2 kW kg^{-1} at a current density of 2.5 A g^{-1}. Paper-based flexible supercapacitor developed by using BFO/graphene nanocomposite as an active electrode material by Soam et al. [130]. In this work, BFO is synthesized by using sol–gel technique. Drop casting method was used to fabricate BFO/graphene on flexible substrate. This kind of flexible substrate demonstrates specific capacitance of 9 mF cm^2 at 10 mV s^{-1}. Golda et al. [131] have synthesized samarium-doped BFO nanoflakes and sheet-like structures applying sol–gel techniques. This kind of doped structure is studied for electrochemical properties. Doped electrode exhibits maximum specific capacitance of 184 F g^{-1}.

The Nath Ghosh [120] et al. have firstly studied and reported excellent cyclic stability along with high-energy density of 18.62 W h kg^{-1} at a power density of 950 W kg^{-1} which makes the BFO–RGO nanocomposite (symmetric device two-electrode system) an excellent electrode material for supercapacitor application. Moreover, they demonstrate a real application of the BFO–RGO as a supercapacitor device, first for charging two symmetric cells, which are connected in series, and this was connected to a 9 V battery for 10 min. After charging the cell, a red light-emitting diode (LED) (1.8 V) was connected and was observed (Fig. 5.8) to successfully light up the LED for 6 min.

5.5 Conclusions

This chapter elaborates the current status and the advancements in the field of supercapacitors on providing requirements essential for better, faster, safer and economic energy storage devices. The principles and energy storage mechanism of the bismuth ferrites and mixed bismuth ferrites have successfully reported attempted. With increasing annealing temperature, the contact angle decreases due to relatively hydrophilic surface formation of BFO electrode material. Hydrophilic surface of the ferrite electrode facilitates several redox reactions for better electrochemical performance. The need for hybrid supercapacitors can be justified due to the limitations

Table 5.1 Some hybrid materials of BFO envisaged previously for supercapacitors

Material	Specific capacitance ($F\,g^{-1}$)	Current density ($A\,g^{-1}$)	Electrolyte	Power density ($W\,kg^{-1}$)	Energy density ($Wh\,kg^{-1}$)	References
BFO–RGO	928.43	5	$3\,M\,KOH + 0.1\,M\,K_4$ [$Fe(CN)_6$]	950	18.62	[120]
$BiFe_{0.95}Cu_{0.05}O_3$	568.13	1	6 M KOH			[121]
5% Ni-doped $BiFeO_3$	513.5	1	6 M KOH			[124]
Perovskite nanocrystalline $BiFeO_3$	81	$1\,mA\,cm^{-2}$	1 M KOH	$3.29\,W\,g^{-1}$	$6.68\,J\,g^{-1}$	[125]
Mixed-phase bismuth ferrite nanoflakes	72	1	2 M NaOH			[126]
$BiFeO_3$ achored TiO_2 nanotube array	440	1.1	$0.5\,M\,Na_2SO_4$	$1.2\,kW\,kg^{-1}$	46.5	[127]
$BiFe_{0.95}Co_{0.05}O_3$	278.2	1	6 M KOH			[128]

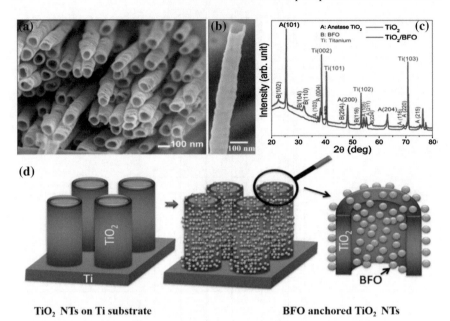

Fig. 5.7 **a**, **b** FE-SEM images of the as-prepared 3D arrays of TiO$_2$/BFO NHs. **c** XRD patterns of the TiO$_2$ NTs and TiO$_2$/BFO NHs. **d** Schematic representation of the anchoring of BFO NPs on the surface of TiO$_2$ NTs [127]

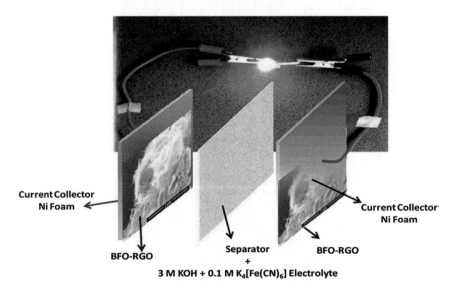

Fig. 5.8 Red light-emitting diode (LED) (1.8 V) powered by two BFO-RGO capacitors connected in series. Glowing LED at different time intervals [120]

of current energy storage devices. The applications of the hybrid supercapacitors are on the rise especially in the field of hybrid energy vehicles. As such, it is important to develop facile supercapacitor materials with various techniques to maximize their usage. Supercapacitors can be sustainable energy storage devices than its contemporaries when synthesized by suitable method on a mass scale on avoiding losses due to resistance and current leakage. The research pertaining to supercapacitors is widely open for exploration and development using bismuth ferrites, and their mixtures which will be of great potential in coming days.

References

1. S. Chu, A. Majumdar, Nature **488**, 294 (2012)
2. M. Winter, R.J. Brodd, Chem. Rev. **104**, 4245 (2004)
3. Y. Zhai, Y. Dou, D. Zhao, P.F. Fulvio, R.T. Mayes, S. Dai, Adv. Mater. **23**, 4828 (2011)
4. X. Lang, A. Hirata, T. Fujita, M. Chen, Nat. Nanotechnol. **6**, 232 (2011)
5. L. Athoue, F. Moser, R. Dugas, O. Crosnier, D. Belanger, T. Brousse, J. Phys. Chem. C **112**, 7270 (2008)
6. J. Yan, E. Khoo, A. Sumboja, P.S. Lee, ACS Nano **4**, 4247 (2010)
7. Z. Chen, V. Augustyn, X. Jia, Q. Xiao, B. Dunn, Y. Lu, ACS Nano **6**, 4319 (2012)
8. Y. Luo, D. Kong, J. Luo, Y. Wang, D. Zhang, K. Qiu, C. Cheng, C. Ming, Li. Ting, RSC Adv. **4**, 13241 (2014)
9. M. Liu, L. Gan, W. Xiong, Z. Xu, D. Zhu, L. Chen, J. Mater. Chem. A **2**, 2555 (2014)
10. M. Sathiya, A.S. Prakash, K. Ramesha, J.M. Tarascon, A.K. Shukla, J. Am. Chem. Soc. **133**, 16291 (2011)
11. A. David, C. Tompsett, C.S. Parker, M.S. Islam, J. Am. Chem. Soc. **136**(4), 1418 (2014)
12. T. Brezesinski, J. Wang, J. Polleux, B. Dunn, S.H. Tolbert, J. Am. Chem. Soc. **131**(5), 1802 (2009)
13. L. Wang, H. Ji, S. Wang, L. Kong, X. Jiang, G. Yang, Nanoscale **5**, 3793 (2013)
14. Y. Liang, M.G. Schwab, L. Zhi, E. Mugnaioli, U. Kolb, X. Feng, K. Mullen, J. Am. Chem. Soc. **132**(42), 15030 (2010)
15. X. Xia, Y. Zhang, D. Chao, C. Guan, Y. Zhang, L. Li, X. Ge, I.M. Bacho, J. Tu, H.J. Fan, Nanoscale **6**, 5008 (2014)
16. J. Kang, A. Hirata, L. Kang, X. Zhang, Y. Hou, L. Chen, C. Li, T. Fujita, K. Akagi, M. Chen, Angew. Chemie **52**, 1664 (2013)
17. Q. Lu, G.J. Chen, J.Q. Xiao, Angew. Chemie **52**, 1882 (2013)
18. J. Petzold, Scripta Mater. **48**, 895 (2003)
19. T.P. Niesen, M.R. De Guire, J. Electroceram. **6**, 169 (2001)
20. M. Yoshimura, W.L. Suchanek, K. Byrappa, Mater. Res. Soc. Bull. **25**, 17 (2000)
21. J. Adam, D. Pendlebury, *Global Research Report: Materials Science and Technology*, vol. 5. (2013), p. 16
22. T.J. Park, Y.B. Mao, S.S. Wong, Chem. Commun. **23**, 22708 (2004)
23. Y.P. Wang, L. Zhou, M.F. Zhang, X.Y. Chen, J.M. Liu, Z.G. Liu, Appl. Phys. Lett. **84**, 1731 (2004)
24. L.W. Martin, Dalton Trans. **39**, 10813 (2010)
25. S.W. Cheong, M. Mostovoy, Nat. Mater. **6**, 13 (2007)
26. M. Winter, R.J. Brodd, Chem. Rev. **104**, 4245 (2005)
27. M.M. Kumar, V.R. Palkar, K. Srinivas, S.V. Suryanarayana, Appl. Phys. Lett. **76**, 1 (2000)
28. V.R. Palkar, R. Pinto, Pramana. J. Phys. **58**, 1003 (2002)
29. N. Adachi, V.P. Denysenkov, S.I. Khartsev, A.M. Grishin, J. Appl. Phys. **88**, 2734 (2000)

30. S. Mohan, B. Subramanian, RSC Adv. **3**, 23737 (2013)
31. R. Das, G.G. Khan, S. Varma, G.D. Mukherjee, K. Mandal, J. Phys. Chem. C **117**(39), 20209 (2013)
32. F. Gao, Y. Yuan, K.F. Wing, X.Y. Chen, F. Chen, J.M. Liu, Z.F. Ren, Appl. Phys. Lett. **89**, 102506 (2006)
33. S.H. Xie, J.Y. Li, R. Proksch, Y.M. Liu, Y.C. Zhou, Y.Y. Liu, L.N. Pan, Y. Qiao, Appl. Phys. Lett. **93**, 222904 (2008)
34. C. Chen, J. Cheng, S. Yu, L. Che, Z. Meng, J. Cryst. Growth **291**, 135 (2006)
35. G. Biasotto, A.Z. Simoes, C.R. Foschini, M.A. Zaghete, J.A. Varela, E. Longo, Mater. Res. Bull. **46**, 2543 (2011)
36. A. Chaudhuri, S. Mitra, M. Mandal, K. Mandal, J. Alloys Compd. **491**, 703 (2010)
37. S.D. Waghmare, V.V. Jadhav, S.K. Gore, S.J. Yoon, S.B. Ambade, B.J. Lokhande, R.S. Mane, S.H. Han, Mater. Res. Bull. **47**, 4169 (2012)
38. S.D. Sartale, C.D. Lokhande, M. Giersig, V. Ganesan, J. Phys.: Condens. Matter **16**, 773 (2004)
39. Y. Mo, M.R. Antonio, D.A. Scherson, J. Phys. Chem. B **104**, 9777 (2000)
40. T.P. Gujar, V.R. Shinde, S.S. Kulkarni, H.M. Pathan, C.D. Lokhande, App. Surf. Science **252**, 3585 (2006)
41. C.D. Lokhande, T.P. Gujar, V.R. Shinde, R.S. Mane, S.H. Han, Electrochem. Comm. **9**, 1805 (2007)
42. N. Dutta, S.K. Bandyopadhyay, S. Rana, P. Sen, A.K. Himanshu, Cornell University Library, arXiv: 1309.5690 (2013)
43. W. Eerenstein, N.D. Mathur, J.F. Scott, Nature **442**, 759 (2006)
44. T.J. Park, G.C. Papaefthymiou, A.J. Viescas, A.R. Moodenbaugh, S.S. Wong, Nano Lett. **7**, 766 (2007)
45. J. Allibe, I.C. Infante, S. Fusil, K. Bouzehouane, E. Jacquet, C. Deranlot, M. Bibes, A. Barthelemy, Appl. Phys. Lett. **95**, 182503 (2009)
46. D. Mazumdar, V. Shelke, M. Iliev, S. Jesse, A. Kumar, S.V. Kalinin, A.P. Baddorf, A. Gupta, Nano Lett. **10**, 2555 (2010)
47. M.K. Singh, Y. Yang, C.G. Takoudis, A. Tatarenko, G. Srinivasan, P. Kharel, G. Lawes, J. Nanosci. Nanotechnol. **10**, 6195 (2010)
48. V. Shelke, D. Mazumdar, G. Srinivasan, A. Kumar, S. Jesse, S. Kalinin, A. Baddorf, A. Gupta, Adv. Mater. **23**, 669 (2011)
49. I.P. Suzdalev, Russ. Chem. Rev. **78**, 249 (2009)
50. G.A. Smolensky, V.A. Isupov, A.I. Agronovskaya, Sov. Phys Solid State **1**, 150 (1959)
51. G. Catalan, J.F. Scott, Adv. Mater. **21**, 2463 (2009)
52. G.D. Achenbach, W.J. James, R. Gerson, Am. J. Ceram Soc. **50**, 437 (1967)
53. F. Kubel, H. Schmid, Acta Crystallogr. B **46**, 698 (1990)
54. J.R. Teague, R. Gerson, W.J. James, Solid State Commun. **8**, 1073 (1970)
55. J. Wang, J.B. Neaton, H. Zheng, V. Nagarajan, S.B. Ogale, B. Liu, D. Viehland, V. Vaithyanathan, D.G. Schlom, U.V. Waghmare, N.A. Spaldin, K.M. Rabe, M. Wuttig, R. Ramesh, Science **299**, 1719 (2003)
56. C. Ederer, N.A. Spaldin, Phys Rev. B **71**, 224103 (2005)
57. J.T. Han, Y.H. Huang, X.J. Wu, C.L. Wu, W. Wei, B. Peng, W. Huang, J.B. Goodenough, Adv. Mater. **18**, 2145 (2006)
58. Y.H. Chu, L.W. Martin, M.B. Holcomb, R. Ramesh, Mater. Today **10**, 16 (2007)
59. F. Gao, X. Chen, K. Yin, S. Dong, Z. Ren, F. Yuan, Adv. Mater. **19**, 2889 (2007)
60. N. Balke, S. Choudhury, S. Jesse, M. Huijben, Y.H. Chu, A.P. Baddorf, L.Q. Chen, R. Ramesh, S.V. Kalinin, Nat. Nanotechnology **4**, 868 (2009)
61. I.M. Sosnowska, J. Microsc. **236**, 109 (2009)
62. Z. Zhang, P. Wu, L. Chen, J. Wang, Appl. Phys. Lett. **96**, 012905 (2010)
63. Z.G. Mei, S. Shang, Y. Wang, Z.K. Liu, Appl. Phys. Lett. **98**, 131904 (2011)
64. S.H. Baek, C.M. Folkman, J.W. Park, S. Lee, C.W. Bark, T. Tybel, Adv. Mater. **23**, 1621 (2011)

65. K.A. McDonnell, N. Wadnerkar, N.J. English, M. Rahman, D. Dowling, Chem. Phys. Lett. **572**, 78 (2013)
66. C. Michel, J.M. Moreau, G.D. Achenbach, R. Gerson, W.J. James, Solid State Commun. **7**, 701 (1969)
67. D. Lebeugle, D. Colson, A. Forget, M. Viret, A.M. Bataille, A. Gukasov, Phys. Rev. Lett. **100**, 227602 (2008)
68. S.Y. Yang, J. Seidel, S.J. Byrnes, P. Shafer, C.H. Yang, M.D. Rossell, P. Yu, Y.H. Chu, J.F. Scott, J.W. Ager, L.W. Martin, R. Ramesh, Nat. Nanotechnol. **5**, 143 (2010)
69. G. Catalan, J. F. Scott, Adv. Mater. 21(24) 2009
70. X. Wang, Y. Zhang, Z. Wu, Mater. Lett. 64 (3) 2010
71. A. Palewicz, I. Sosnowska, R. Przenios, A.W. Hewat, Acta Phys. Pol., A **117**, 296 (2010)
72. J.F. Scott, J. Magn. Magn. Mater. **321**, 1689 (2009)
73. J.G. Ismilzade, Phys. Status Solidi (b) **46**, K39 (1971)
74. G.A. Smolenskii, V.M. Yudin, Sov. Phys. JETP **16**, 622 (1963)
75. X.M. Sun, Y.D. Li, J. Eur. Chem. **9**, 2229 (2003)
76. B.E. Conway, W.G. Pell, J. Power Sources **105**, 169 (2002)
77. F. Tao, Y.Q. Zhao, G.Q. Zhang, H.L. Li, Electrochem. Commun. **9**, 1282 (2007)
78. Y. Li, K. Huang, D. Zeng, S. Liu, Z. Yao, J. Solid State Electrochem. **14**, 1205 (2010)
79. H. Schmid, Ferroelectrics **162**, 317 (1994)
80. J. Ryu, S. Priya, K. Uchino, H.J. Kim, Phys. J. Electroceramics **8**, 112 (2002)
81. M. Vopsaroiu, J. Blackburn, M.G. Cain, J. Phys. D Appl. Phys. **40**, 5027 (2007)
82. J.P. Rivera, Ferroelectrics **161**, 165 (1994)
83. B.I. AlShin, D.N. Astrov, So. Phys. JETP **17**, 809 (1963)
84. G.T. Rado, Phys. Rev. Lett. **13**, 335 (1964)
85. T. Watanabe, K. Kohn, Phase Trans. **15**, 57 (1989)
86. N.A. Hill, J. Phys. Chem. B **104**, 6694 (2000)
87. M. Fiebig, T. Lottermoser, D. Frohlich, A.V. Goltsev, R.V. Pisarev, Nature **419**, 818 (2002)
88. V. Srivastava, Y. Song, K. Bhatti, R.D. James, Adv. Energy Mater. **1**, 97 (2011)
89. R.M. Bozorth, E.F. Tilden, A. Williams, J. Phys. Rev. **99**, 1788 (1955)
90. F. Kubel, H. Schmid, Acta Crystallogr. Sect. B: Struct. Sci. **46**, 698 (1990)
91. N.M. Aimon, D.H. Kim, H.K. Choi, C.A. Ross, Appl. Phys. Let. **100**, 092901 (2012)
92. P. Calvani, M. Capizzi, F. Donato, S. Lupi, P. Maselli, D. Peschiaroli, Phys. Rev. B **47**, 8917 (1993)
93. J. Ma, J. Hu, Z. Li, C.W. Nan, Adv. Mater. **23**, 1062 (2011)
94. H. Zheng, J. Wang, S. E. Lofland, Z. Ma, L. Mohaddes-Aradabili, T. Zhao, L. Salamanca-Riba, S.R. Shinde, S.B. Ogale, F. Bai, D. Viehland, Y. Jia, D.G. Schlom, M. Wuttig, A. Roytburd, R. Ramesh, Science **303**, 661 (2004)
95. R. Comes, H. Liu, M. Kholkhov, R. Kasica, J. Lu, S.A. Wolf, Nano Lett. **12**, 2367 (2012)
96. X. Liu, Y. Kim, S. Goetze, X. Li, S. Dong, P. Werner, M. Alexe, D. Hesse, Nano Lett. **11**, 3202 (2012)
97. G. Liu, C.W. Nan, N. Cai, Y. Lin, J. Applied Physics **95**, 2660 (2004)
98. J. Zhai, N. Cai, Z. Shi, Y. Lin, C.W. Nan, J. Appl. Phys. **95**, 5685 (2004)
99. J. Van Suchtelen, Philips Res. Rep. **27**, 28 (1972)
100. I. Levin, J. Li, J. Slutsker, A. Roytburd, Adv. Mater. **18**, 2044 (2006)
101. Q. Zhan, R. Yu, S.P. Crane, H. Zheng, C. Kisielowski, R. Ramesh, Appl. Phys. Lett. **89**, 172902 (2006)
102. F. Zavaliche, H. Zheng, L. Mohaddes-Ardabili, S.Y. Yang, Q. Zhan, P. Shafer, E. Reilly, R. Chopdekar, Y. Jia, P. Wright, D.G. Schlom, Y. Suzuki, R. Ramesh, Nano Lett. **5**, 1793 (2005)
103. H. Zheng, F. Straub, Q. Zhan, P.L. Yang, W.K. Hsieh, F. Zavaliche, Y.H. Chu, U. Dahmen, R. Ramesh, Adv. Mater. **18**, 2747 (2006)
104. S.M. Stratulat, X. Lu, A. Morelli, D. Hesse, W. Erfurth, M. Alexe, Nano Lett. **13**, 3884 (2013)
105. T. Rojac, A. Bencan, B. Malic, G. Tutuncu, J.L. Jones, J.E. Daniels, D. Damjanovic, J. Am. Ceram. Soc. **97**, 1993 (2014)
106. Y. Wei, X. Wang, J. Zhu, X. Wang, J. Jia, J. Am. Ceram. Soc. **96**, 3163 (2013)

107. A. Mishra, B. Majumdar, R. Ranjan, J. Eur. Ceram. Soc. **37**, 2379 (2017)
108. S. Sharma, V. Singh, R.K. Dwivedi, J. Alloy. Compd. **682**, 723 (2016)
109. T.M. Correia, M. McMillen, M.K. Rokosz, P.M. Weaver, J.M. Gregg, G. Viola, M.G. Cain, J. Am. Ceram. Soc. **96**, 2699 (2013)
110. H. Pan, Y. Zeng, Y. Shen, Y.H. Lin, J. Ma, L. Li, C.W. Nan, J. Mater. Chem. A **5**, 5920 (2017)
111. H. Pan, J. Ma, J. Ma, Q. Zhang, X. Liu, B. Guan, L. Gu, X. Zhang, Y.J. Zhang, L. Li, Y. Shen, Nat. Commun **9**, 1 (2018)
112. B. Xu, J. Iniguez, L. Bellaiche, Nat. Commun. **8**, 15682 (2017)
113. L. Li, H. Bi, S. Gai, F. He, P. Gao, Y. Dai, X. Zhang, D. Yang, M. Zhang, P. Yang, Sci. Rep. **7**, 43116 (2017)
114. W. Zhang, B. Quan, C. Lee, S.K. Park, X. Li, E. Choi, G. Diao, Y. Piao, A.C.S. Appl, Mater. Interfaces **7**, 2404 (2015)
115. Z. Wang, X. Zhang, Y. Li, Z. Liu, Z. Hao, J. Mater. Chem. A **21**, 6393 (2013)
116. W. Cai, T. Lai, W. Dai, J. Ye, J. Power Sources **255**, 170 (2014)
117. P. Xiong, H. Huang, X. Wang, J. Power Sources **245**, 937 (2014)
118. C. Xiang, M. Li, M. Zhi, A. Manivannan, N. Wu, J. Power Sources **226**, 65 (2013)
119. P. Xiong, C. Hu, Y. Fan, W. Zhang, J. Zhu, X. Wang, J. Power Sources **266**, 384 (2014)
120. D. Moitra, C. Anand, B.K. Ghosh, M. Chandel, N.N., Ghosh, ACS Appl. Energy Mater. **1**, 464 (2018)
121. J. Khajonrit, S. Phumying, S. Maensiri, Jpn. J. Appl. Phys. **55**, 06GJ14 (2016)
122. H. Heli, H. Yadegari, J. Electroanalytical Chem. **713**, 103 (2014)
123. R. Rakhi, W. Chen, D. Cha, H.N. Alshareef, J. Mater. Chem. **21**, 16197 (2011)
124. J. Khajonrit, N. Prasoetsopha, T. Sinprachim, P. Kidkhunthod, S. Pinitsoontorn, S. Maensiri, Adv. Nat. Sci.: Nanosci. Nanotechnol. **8**, 015010 (2017)
125. C. Lokhande, T. Gujar, V. Shinde, R.S. Mane, S.H. Han, Electrochem. Commun. **9**, 1805 (2007)
126. V.V. Jadhav, M.K. Zate, S. Liu, M. Naushad, R.S. Mane, K. Hui, S.H. Han, Appl. Nanosci. **6**, 511 (2016)
127. A. Sarkar, A.K. Singh, D. Sarkar, G.G. Khan, K. Mandal, ACS Sustainable Chem. Eng. **3**, 2254 (2015)
128. J. Khajonrit, U. Wongpratat, P. Kidkhunthod, S. Pinitsoontorn, S. Maensiri, J. Magn. Magn. Mater. **449**, 423 (2018)
129. S. Nayak, A. Soam, J. Nanda, C. Mahender, M. Singh, D. Mohapatra, R. Kumar, J. Mater. Sci.: Mater. Electronics **29**, 9361–9368 (2018)
130. A. Soam, R. Kumar, C. Mahender, M. Singh, D. Thatoi, R.O. Dusane, J. Alloy. Comp. **813**, 152145 (2020)
131. R.A. Golda, A. Marikani, E.J. Alex, Ceram. Int. **46**, 1962 (2020)

Chapter 6
Limitations and Perspectives

With the impending energy crisis, this and next few human generations need to look at alternative renewable energy sources for fulfilling future needs. The utilization of energy from renewable sources depends on our ability to store the generated energy in electrochemical charge storage devices and to retrieve when needed. Thus, in recent times there is an impetus to the research on electric energy storage devices such as electrochemical supercapacitor (ESs) and batteries. Also, due to the fast-growing market for portable electronic devices and the development of hybrid electric vehicles, there has been an ever-increasing and urgent need for environmentally friendly high-power energy storage resources. For example, in major cities of India, the pollution level has already been reached the above level of acceptance.

Recently, ES-based technology has received much attention due to the potential of producing high-power energy storage devices. Research and development of ES devices have gained importance in recent years. There are myriad applications of supercapacitors that are not limited to; (a) mobile phones, (b) laptops, (c) wireless communication devices, (d) uninterrupted power supply systems, (e) missile technologies, (f) hybrid and electric vehicles, (g) notebook computers, (h) toys, (i) automotive auxiliary devices, (j) automotive engine stating devices, etc. There is a need to fabricate miniaturized portable ES devices with high power and energy which can fill the gap between the batteries and conventional capacitors. In general, ES materials are tested through cyclic voltammetry and galvanostatic charge–discharge measurements. Many transition metal oxides like ZnO, TiO_2, RuO_2, NiO, Co_3O_4, MoO_3, MnO_2, SnO_2, ZnO, TiO_2, V_2O_5, CuO, Fe_2O_3, WO_3, etc., have already been envisaged as supercapacitor electrode materials for higher and durable electrochemical energy storage performance.

The forthcoming of supercapacitors lies in the optimization of present materials to encounter the mandate of high-energy density while continuing the high cycle life, lower charging time, for exchanging the rechargeable batteries in mobile phone

© The Authors 2020
V. V. Jadhav et al., *Bismuth-Ferrite-Based Electrochemical Supercapacitors*,
SpringerBriefs in Materials, https://doi.org/10.1007/978-3-030-16718-9_6

industry and in many further applications. The current lithium-ion battery stores high-energy density but the issue is with safety. Supercapacitors are safer than lithium-ion battery with low internal resistance and lower heat in short circuit. Therefore, it could be an alternative energy storage device which can replace batteries with high power and energy density with lightweight. Hence, nowadays, supercapacitor market is growing very rapidly in hybrid electric cars.

Ferrites, in addition to special magnetic properties, have attracted great attention in energy storage devices. Also, today's electronics industry is looking for tiny and miniaturized devices with low-cost new materials. In the field of smart materials, multiferroic materials are at the top of research in materials science. Their fascinating applications include antenna rod, transformer core, recording head, loading coil, memory, and microwave devices, etc. However, very fewer reports are available on their use in electrochemical energy storage devices. Various physical, as well as chemical methods, are being used for the synthesis of ferrite thin-film electrodes. Solution-processed methods have found great advantages due to their low-cost, easy handling potential, non-expensive instruments, and easy availability. They demonstrate the capability of controlling composition and morphology by varying electrochemical parameters and the ability to deposit films on a complex surface. They also are non-vacuum and suitable methods to prepare electrodes of large surface area and can be used for the preparation of both thin- and thick-film electrodes.

6.1 Limitations

In this book, we summarized recent advances in BFO and BFO-based materials for energy storage applications. In order to realize better electrochemical performance, researchers have developed novel morphologies, structures, and composites of BFO and BFO-based materials for high energy and power. However, there is still space to optimization of BFO and BFO-based electrode materials for better performance. Few of these are as follows;

1. Previous studies identified the critical role of morphology and structure of electrode materials while obtaining moderate electrochemical performance. Novel structures need to be synthesized for better and sustainable electrochemical performance and chemical stability. It is noteworthy to mention that nanostructured materials should have a high specific area and more active sites, and novel 3D micron or submicron-structured electrode materials need to be in electrochemical supercapacitors.

2. Till date, more work has been focused on electrode materials rather than electrolyte solutions, separators, and counter electrodes in three-electrode system. The future works should be focused to optimize all involved components for better performance and practical potential.

3. More works need to be emphasized on fabricating novel morphologies and structures, with clear mechanism responsible for charging–discharging activity during
   energy storage process.
4. Current applied approaches to fabricate BFO and BFO-based materials are limited to classical methods such as hydrothermal, solvothermal, and sol–gel, which
   are responsible for impurity involvement into electrode materials which later can
   adversely effect on electrochemical performance. Developing novel defect-free
   method for ferrite nanostructures of various morphologies and surface area would
   be fascinating.
5. The scale-up preparation of BFO, BFO-based, and other metal oxide electrode
   materials is still challenging for commercial benefits.
6. Adaptability to the load change and durability under severe working conditions
   of assembled device should be tested. The current gap implied a long way ahead
   toward the commercialization.

6.2 Perspectives

The ferrites, specially bismuth ferrite ($BiFeO_3$–BFO), due to multifunctional properties with Currie temperature (T_c) close to 1103 K, proved their worth in various
applications such as memory, spintronics and magnetoelectric sensors, photovoltaic
devices, capacitors, fuel cells, solar cells, batteries, and photoelectrochemical water
splitting [1–7]. Among these applications, BFO shows excellent electrochemical
properties which can be the good alternative to renewable energy sources.

The electrochemical properties of the BFO have various aspects which can affect
the performance such as synthesis method, morphology, suitable dopant, surface
area, and conductivity. The surface area of the nanosized BFO plays an important
role in the electrochemical performance. Pristine BFO has comparatively small surface area which can be improved using various dopants. The graphene/carbonaceous
materials are the materials with high surface area. The adequate composition of these
materials can surely improve the performance the electrochemical performance of
the BFO electrode. The conductivity of the electrode material is also an important
factor which is necessary for the betterment in the electrochemical performance of
the BFO electrode material. Nayak et al. prepared $BiFeO_3$-graphene nanocomposite
using sol–gel autocombustion method for supercapacitor application. A capacitance
of the pristine BFO was found ~5.7 mF/cm^2 but after adding graphene sheets capacitance was increased to ~11.35 mF/cm^2. The capacitance of BFO electrode per unit
mass was found to be 64 F/g @ 20 mV/s in Na_2SO_4 which is analogous with the
previous reports. The performance of the electrode material is also depending on the
electrolyte. The electrochemical performance of the BFO electrode was tested in various electrolytes such as NaOH, KOH, Na_2SO_3, Na_2SO_4, NaCl, and KCl. Lokhande
et al. observed that BFO shows excellent electrochemical properties in NaOH electrolyte [8, 9]. Sarkar et al. reported the TiO_2/$BiFeO_3$ (NH) nanoheterostructure array

and its electrochemical properties. The specific capacitance of the NHs was found to be 440 F/g @ a current density of 1.1 A/g with good energy and power density, i.e., 46.5 Wh/kg and 1.2 kW/kg, respectively, @ current density of 2.5 A/g [10]. The specific capacitance of the $TiO_2/BiFeO_3$ NHs was found much better than the previously reported carbon-based electrode materials such as carbon nanotubes with graphene (~199 F/g) [11] and graphene/carbon nanotube (115 F/g) [12]. The morphology of the electrode material also plays an important role in the electrochemical properties. Jadhav et al. studied the electrochemical properties of the nanoflakes of bismuth ferrite using three-electrode system. The effect of electrolyte concentration was also studied for 0.25, 0.5, 0.75, 1, 1.5, and 2 M NaOH electrolyte @10 mV/s. The specific capacitance of BFO was found to be 101.63 F/g for 10 mV scan rate. Authors also observed that with increasing scan rate, the specific capacitance was decreased which shows states that more active sites are available at low scan rates. The specific capacitance of BFO electrode determined using galvanostatic charge–discharge (GCD) was found 72.2 F/g @ 1 A/g in 2 M NaOH solution [13]. Lin et al., studied the effect of etching on the morphology and electrochemical properties of the $BiFeO_3$ and $BiFeO_3$-based electrodes. The etching time was varied for 30–90 min. The highest specific capacitance of 561.48 F/g was found at the scan rate of 2 mV/s for etching time 60 min with god cyclic stability after 1500 cycles @ current density of 5 A/g. The morphology-based electrochemical performance of the $BiFeO_3$ and $BiFeO_3$-based electrodes was studied in various electrolytes. The specific capacitance of the $BiFeO_3$ thin film was found to be 81 F/g in 1 M NaOH but $BiFeO_3$ nanoparticles show the much better specific capacitance of 342.97 F/g in 6 M KOH. Again, the $Bi_{0.9}La_{0.1}FeO_3$ (BLFO) reveals higher specific capacitance of 561.48 in 1 M NaOH electrolyte which shows the porosity effect of BLFO microsphere [14]. This study attributed to the nanoparticle morphology supports the fast charge transfer and electrolyte ions diffusion [15, 16]. The electrochemical properties of the electrode materials are also depending on surface area. The high surface area of the electrode material is attributed to the increasing electrochemical properties. The surface area of the electrode materials can be improved by various techniques such as doping of suitable dopant, using surfactant material, by adjusting calcination temperature. Moitra et al., reported the nanocomposite of $BiFeO_3$ nanowire with reduced graphene oxide (BFO-RGO) using hydrothermal method which results into increase in surface area and hence the specific capacitance of the prepared electrode [17]. The specific capacitance of the BFO-RGO was found to be 928.43 F/g @ 5 A/g which is much high than pristine BFO electrode (331.71 F/g). The specific capacitance was decreased with increased scan rate which may be due to decrease in diffusion rate of OH⁻ into electrode surface. The charge and power density (18.62 W and 950 W/kg, respectively) of the BFO–RGO also found high as compared with pristine BFO.

6.3 Directions

1. The work presented in this book has suggested the plausible uses of the bismuth ferrites and their composts for supercapacitor applications.
2. Also, one can use ferrite-based perovskites for electrochemical supercapacitor applications. It would be a wonderful choice if these structures are mixed with perovskites or MXenes for obtaining better supercapacitor performance and stability.
3. Developing BFO composites with N-doped carbon materials, on conducting polymers, and metal oxides for better electrochemical supercapacitor performance will open different opportunity.
4. Finally, BFO and its composite-based either powder or thin-film-type electrode materials can pave the new and promising way to develop flexible, lightweight, and portable electronic devices for wearable applications.

References

1. M. Dziubaniuk, R. Bujakiewicz-Korońska, J. Suchanicz, J. Wyrwa, Mieczysław Rękas, Application of bismuth ferrite protonic conductor for ammonia gas detection. Sens. Actuators B Chem. **188**, 957–964 (2013)
2. L. Fei, Y. Hu, X. Li, R. Song, L. Sun, H. Huang, H. Gu, H.L.W. Chan, Y. Wang, Electrospun bismuth ferrite nanofibers for potential applications in ferroelectric photovoltaic devices. ACS Appl. Mater. Interfaces **7**(6), 3665–3670 (2015)
3. Y. Lei, J. Li, Y. Wang, L. Gu, Y. Chang, H. Yuan, D. Xiao, Rapid microwave-assisted green synthesis of 3D hierarchical flower-shaped $NiCo_2O_4$ microsphere for high-performance supercapacitor. ACS Appl. Mater. Interfaces **6**(3), 1773–1780 (2014)
4. A. Jayakumar, R.P. Antony, R. Wang, J.M. Lee, MOF-derived hollow cage $Ni_xCo_{3-x}O_4$ and their synergy with graphene for outstanding supercapacitors. Small **13**(11), 1603102 (2017)
5. F. Zheng, D. Zhu, Q. Chen, Facile fabrication of porous $Ni_xCo_{3-x}O_4$ nanosheets with enhanced electrochemical performance as anode materials for Li-ion batteries. ACS Appl. Mater. Interfaces **6**(12), 9256–9264 (2014)
6. Z. Yin, Q. Zheng, S.C. Chen, D. Cai, Interface control of semiconducting metal oxide layers for efficient and stable inverted polymer solar cells with open-circuit voltages over 1.0 volt. ACS Appl. Mater. Interfaces **5**(18), 9015–9025 (2013)
7. T. Soltani, A. Tayyebi, B.K. Lee, $BiFeO_3/BiVO_4$ p-n heterojunction for efficient and stable photocatalytic and photoelectrochemical water splitting under visible-light irradiation. Catal. Today **340**, 188–196 (2020)
8. S. Nayak, J. Ankur Soam, C. Nanda, M. Mamraj Singh, D. Mohapatra, R. Kumar, Sol–gel synthesized $BiFeO_3$–graphene nanocomposite as efficient electrode for supercapacitor application. J. Mater. Sci. Mater. Electron. **29**(11), 9361–9368 (2018)
9. C.D. Lokhande, T.P. Gujar, V.R. Shinde, R.S. Mane, S.-H. Han, Electrochemical supercapacitor application of pervoskite thin films. Electrochem. Commun. **9**(7), 1805–1809 (2007)
10. A. Sarkar, A.K. Singh, D. Sarkar, G.G. Khan, K. Mandal, Three-dimensional nanoarchitecture of $BiFeO_3$ anchored TiO_2 nanotube arrays for electrochemical energy storage and solar energy conversion. ACS Sustain. Chem. Eng. **3**(9), 2254–2263 (2015)
11. D. T. Pham, T.H. Lee, D.H. Luong, F. Yao, A. Ghosh, V.T. Le, T.H. Kim, B. Li, J. Chang, Y.H. Lee, Carbon nanotube-bridged graphene 3D building blocks for ultrafast compact supercapacitors. ACS Nano **9**(2), 2018–2027 (2015)

12. M. Notarianni, J. Liu, F. Mirri, M. Pasquali, N. Motta, Graphene-based supercapacitor with carbon nanotube film as highly efficient current collector. Nanotechnology **25**(43), 435405 (2014)
13. V.V. Jadhav, M.K. Zate, S. Liu, M. Naushad, R.S. Mane, K.N. Hui, S.-H. Han, Mixed-phase bismuth ferrite nanoflake electrodes for supercapacitor application. Appl. Nanosci. **6**(4), 511–519 (2016)
14. Y. Lin, Q. Wang, S. Gao, H. Yang, P. Liu, Morphology-controlled porous $Bi_{0.9}La_{0.1}FeO_3$ microspheres for applications in supercapacitors. Ceram. Int. **44**(3), 2649–2655 (2018)
15. C.D. Lokhande, D.P. Dubal, O.-S. Joo, Metal oxide thin film-based supercapacitors. Curr. Appl. Phys. **11**(3), 255–270 (2011)
16. S.K. Bandyopadhyay, N. Dutta, S. Rana, P. Sen, A.K. Himanshu, P.K. Chakraborty, High value of ferroelectric polarization in BFO nanorods, in *AIP Conference Proceedings,* vol. 1591, no. 1, pp. 1680–1682. American Institute of Physics (2014)
17. D. Moitra, C. Anand, B.K. Ghosh, M. Chandel, N.N. Ghosh, One-dimensional $BiFeO_3$ nanowire-reduced graphene oxide nanocomposite as excellent supercapacitor electrode material. ACS Appl. Energy Mater. **1**(2), 464–474 (2018)

Printed in the United States
By Bookmasters